Correlation of USA Daily Math Grade 2 to Common Core State Standards for Mathematics

2.OA	Operations and Algebraic Thinking
2.OA.1 Use addition and subtraction within 100 to solve one- and two-step word problems with unknowns in all positions.	**Mondays** p. 19 #2 p. 21 #4 p. 31 #2 p. 34 #2 p. 60 #2 p. 70 #2 p. 76 #3 p. 85 #4 p. 88 #3 **Thursdays** p. 20 #2 p. 23 #2 **Fridays** p. 27 #3 p. 30 #3–4 p. 33 #2, 4 p. 40 #3 p. 75 #2 **Brain Stretch** pp. 3, 6, 9, 12, 15, 18, 60, 63, 66, 69, 72, 75, 78, 81, 84
2.OA.2 Fluently add and subtract within 20 using mental strategies.	**Mondays** p. 1 #2 p. 4 #2 p. 7 #2 p. 19 #3 p. 22 #3 p. 25 #2 p. 28 #2, 4 p. 31 #1 p. 34 #1, 4 p. 37 #1 p. 43 #1, 3 p. 46 #1, 3 p. 49 #1, 3 p. 52 #1 p. 55 #1 p. 58 #1 p. 61 #1 p. 64 #1 p. 67 #1–2 p. 79 #4 p. 82 #4 p. 88 #1 **Tuesdays** p. 1 #3 p. 4 #3 p. 19 #2 p. 22 #2 p. 28 #2 p. 46 #3 p. 49 #2 p. 55 #2 p. 58 #2 p. 67 #4 p. 79 #4 p. 82 #3 p. 85 #1 **Friday** p. 24 #3–4 p. 87 #3 **Brain Stretch** pp. 21, 24, 30, 33, 36, 67
2.OA.3 Determine whether a group of objects has an odd or even number of members; write an equation to express an even number as a sum of two equal addends.	**Tuesday** p. 31 #2 p. 40 #2 p. 43 #3
2.OA.4 Use addition to find the total number of objects arranged in rectangular arrays; write an equation to express the total as a sum of equal addends.	**Monday** p. 37 #2 p. 46 #4 **Tuesday** p. 76 #2 p. 79 #1 p. 82 #1
2.NBT	**Number and Operations in Base Ten**
2.NBT.1 Understand place value.	**Tuesdays** p. 1 #1 p. 4 #1 p. 7 #1–2 p. 10 #1 p. 13 #1 p. 16 #1, 3 p. 19 #1 p. 22 #1 p. 25 #1 p. 28 #1, 3 p. 31 #1 p. 34 #1 p. 37 #1 p. 40 #1 p. 43 #1–2 p. 46 #1 p. 49 #1, 3 p. 52 #1, 3 p. 55 #1
2.NBT.2 Count within 1000; skip-count by 5s, 10s, and 100s.	**Mondays** p. 4 #3 p. 7 #3 p. 13 #1–2 p. 16 #1 p. 22 #1 p. 76 #1 p. 79 #1, 3 **Tuesdays** p. 1 #3 p. 22 #4
2.NBT.3 Read and write numbers to 1000 using base-ten numerals, number names, and expanded form.	**Monday** p. 28 #3 **Tuesdays** p. 1 #2 p. 4 #2 p. 28 #3 p. 37 #3 p. 49 #3 p. 55 #3 p. 70 #4 p. 85 #2 p. 88 #2
2.NBT.4 Compare two three-digit numbers based on meanings of the hundreds, tens, and ones digits, using >, =, and <.	**Tuesdays** p. 25 #2 p. 34 #2 p. 37 #2 p. 40 #3 p. 67 #1 p. 70 #1
2.NBT.5 Fluently add and subtract within 100 using strategies.	**Mondays** p. 70 #1–2 p. 73 #1–2 p. 76 #2 p. 79 #2 p. 82 #1–2 p. 85 #1–2 **Tuesdays** p. 73 #3 p. 88 #3 **Friday** p. 72 #4 **Brain Stretch** pp. 12, 27, 30, 33, 36, 39, 42, 45, 48, 51, 60, 72, 75, 78, 81, 87
2.NBT.6 Add up to four two-digit numbers using strategies based on place value and properties of operations.	**Tuesdays** p. 85 #4 p. 88 #4
2.NBT.7 Add and subtract within 1000, using concrete models or strategies.	**Tuesdays** p. 19 #2 p. 22 #2 p. 25 #4 p. 28 #4 p. 31 #4 p. 34 #4 p. 37 #4 p. 40 #4 p. 43 #4 p. 61 #3 p. 64 #2–3 p. 67 #3 p. 73 #1, 3 p. 76 #1, 3 p. 79 #3 p. 85 #3 **Brain Stretch** pp. 54, 57
2.NBT.8 Mentally add or subtract 10 or 100 to a given number 100–900.	**Mondays** p. 34 #3 p. 67 #4 p. 73 #3 p. 82 #3 **Tuesdays** p. 13 #4 p. 16 #2 p. 19 #4 p. 25 #3 p. 34 #3 p. 52 #2
2.NBT.9 Explain why addition and subtraction strategies work, using place value and the properties of operations.	**Tuesdays** p. 67 #2 p. 70 #3

2.MD	Measurement and Data
2.MD.1 Measure the length of an object by selecting and using appropriate tools.	
2.MD.2 Measure the length of an object twice, using length units of different lengths for the two measurements; describe how the two measurements relate to the size of the unit chosen.	
2.MD.3 Estimate lengths using units of inches, feet, centimeters, and meters.	**Thursdays** p. 11 #3 p. 32 #4 p. 38 #2, 4 p. 47 #2 p. 59 #3 p. 62 #3 p. 68 #2
2.MD.4 Measure to determine how much longer one object is than another.	
2.MD.5 Use addition and subtraction within 100 to solve word problems involving lengths.	**Thursdays** p. 53 #2 p. 59 #4
2.MD.6 Represent whole numbers (and sums and differences) as lengths from 0 on a number line.	**Tuesday** p. 43 #2 **Thursdays** p. 44 #2 p. 56 #3
2.MD.7 Tell and write time from analog and digital clocks to the nearest five minutes, using a.m. and p.m.	**Thursdays** p. 2 #1 p. 8 #1 p. 11 #1 p. 14 #1 p. 20 #1 p. 23 #1 p. 26 #1 p. 29 #1–2 p. 32 #1 p. 35 #1 p. 38 #1 p. 41 #1 p. 47 #1 p. 50 #1 p. 53 #1 p. 56 #1 p. 59 #1 p. 62 #1 p. 65 #1 p. 68 #1 p. 74 #1 p. 77 #1 p. 80 #1 p. 83 #1 p. 86 #1 p. 89 #1
2.MD.8 Solve word problems involving dollar bills, quarters, dimes, nickels, and pennies, using $ and ¢ symbols appropriately.	**Tuesdays** p. 19 #2 p. 22 #2 p. 25 #4 p. 28 #4 p. 31 #4 p. 34 #4 p. 37 #4 p. 40 #4 p. 43 #4 **Thursdays** p. 86 #3 p. 89 #3 **Brain Stretch** p. 90
2.MD.9 Generate measurement data. Show the measurements by making a line plot.	
2.MD.10 Draw a picture graph and a bar graph to represent a data set. Solve simple problems using a bar graph.	**Fridays** p. 3 #1–4 p. 6 #1–4 p. 9 #1–4 p. 12 #1–4 p. 15 #1–4 p. 18 #1–4 p. 21 #1–5 p. 24 #1–4 p. 27 #1–4 p. 30 #1–4 p. 33 #1–4 p. 39 #1–4 p. 42 #1–4 p. 45 #1–4 p. 60 #1–4 p. 63 #1–4 p. 66 #1–4 p. 69 #1–5 p. 72 #1–5 p. 75 #1–3 p. 78 #1–4 p. 81 #1–4 p. 87 #1–3
2.G	**Geometry**
2.G.1 Recognize and draw shapes having specified attributes. Identify triangles, quadrilaterals, pentagons, hexagons, and cubes.	**Wednesdays** p. 2 #1–3 p. 5 #1–3 p. 8 #1–3 p. 11 #1–3 p. 14 #1–3 p. 17 #1–3 p. 26 #1–3 p. 32 #1–3, 5 p. 35 #4 p. 38 #4–5 p. 41 #4–5 p. 44 #4–5 p. 47 #2–3 p. 50 #1–3 p. 53 #1 p. 56 #1 p. 59 #1 p. 62 #1 p. 65 #1 p. 68 #1 p. 71 #1 p. 74 #1 p. 80 #1–2 p. 83 #1 p. 86 #1, 3 p. 89 #1, 3
2.G.2 Partition a rectangle into rows and columns of same-size squares and count to find the total number of them.	**Thursdays** p. 65 #3–4 p. 68 #3–4 p. 71 #4 p. 74 #3
2.G.3 Partition circles and rectangles into two, three, or four equal shares, describe the shares using words.	**Thursdays** p. 71 #3 p. 74 #4 p. 80 #3–4 p. 89 #4

MONDAY Patterning and Algebra

1. Color the shapes to make a pattern.

What is your pattern rule? ______________________________

2. What is the missing number?

_______ + 8 = 12

3. Count on by 2s from 20.

20, ______, ______, ______

TUESDAY Number Sense and Operations

1. What is the number?

tens ______

ones ______

number ______

2. Write the numeral.

sixteen _________

3. Add.

1 + 6 + 3 = ________

4. What is the name of this coin?

A. nickel

B. quarter

C. penny

WEDNESDAY Geometry

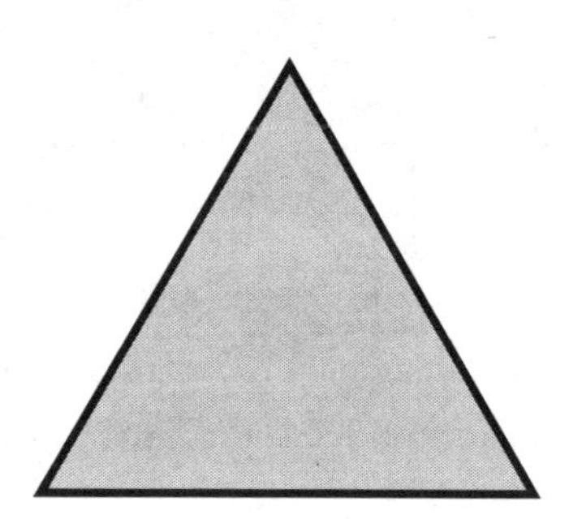

1. Circle the name of this shape.

rectangle triangle

2. How many sides does it have? ________

3. How many vertices does it have? ________

Trace and draw the shape.

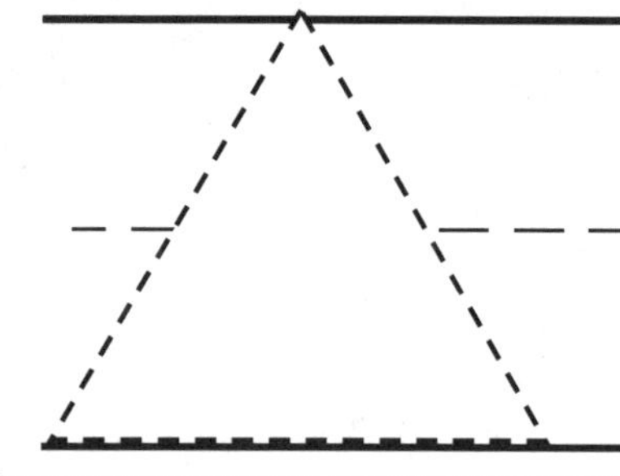

THURSDAY Measurement

1. What time is it?

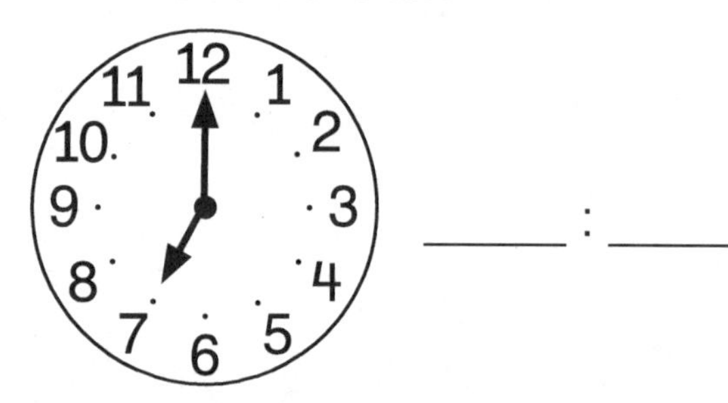

____ : ____

2. Is the temperature hot or cold?

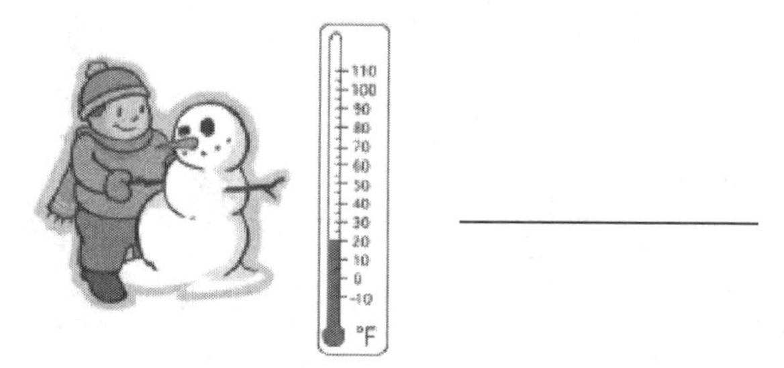

____

3. Which container holds more?

A. 

B. 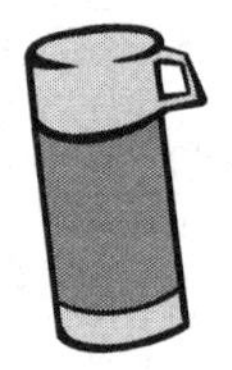

4. Measure the length of the line.

It is about ________ long.

FRIDAY Data Management

Mr. Tate's students conducted a survey of their favorite pets.
Use the pictograph to answer the questions about the results.

Favorite Pets

1. How many students chose  ? ____________

2. How many students chose  ? ____________

3. Circle the pet students chose the most.   

4. Circle the pet students chose the least.   

BRAIN STRETCH

The clown had 5 red balloons and 4 blue balloons.
How many balloons did the clown have in all?

MONDAY Patterning and Algebra

1. Color the shapes to make a pattern.

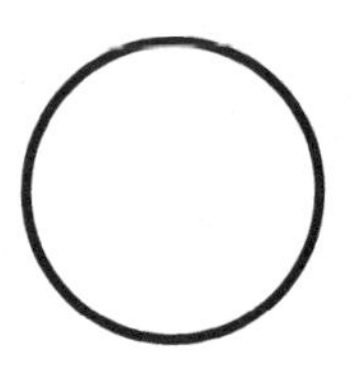 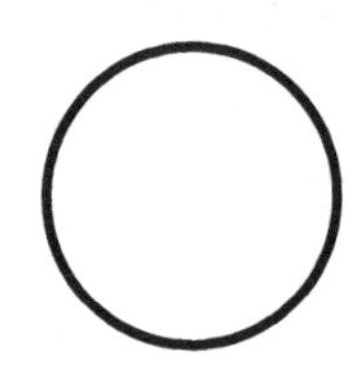 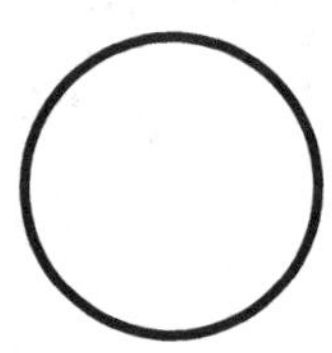 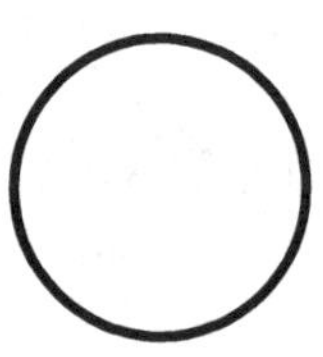 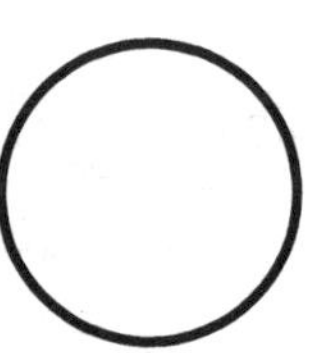 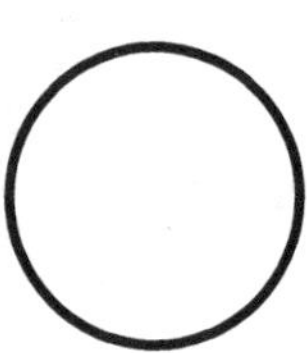

What is your pattern rule? ______________________________

2. What is the missing number?

_______ + 4 = 18

3. Count on by 10s from 70.

70, ________, ________, ________

TUESDAY Number Sense and Operations

1. What is the number?

tens ______

ones ______

number ______

2. Write the numeral.

A. forty ___________

B. thirteen ___________

3. Add.

6 + 7 + 2 = ________

4. What is the name of this coin?

A. nickel
B. quarter
C. penny

WEDNESDAY Geometry

1. Circle the name of this shape.

   circle   triangle

2. How many sides does it have? _______

3. How many vertices does it have? _______

Trace and draw the shape.

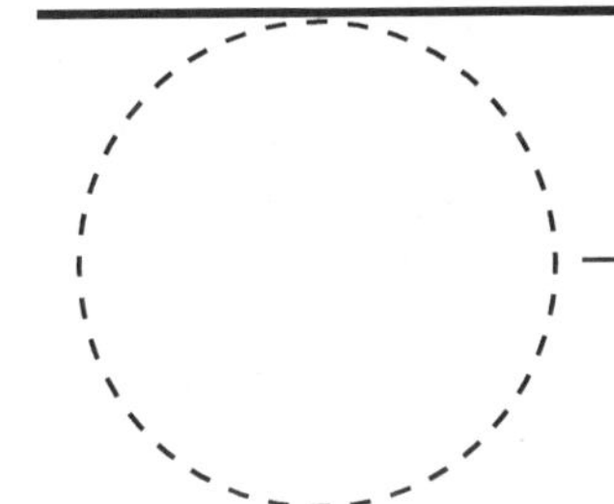

THURSDAY Measurement

1. Which tool would be best to measure the length of a book?

   A. scale

   B. measuring tape

   C. clock

2. Choose the better unit of measure for the weight of a cat.

   A. gram

   B. kilogram

3. Which container holds less?

   A.    B. 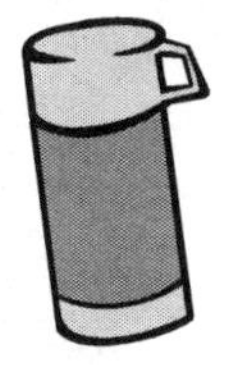

4. Measure the length of the line.

   It is about _______  long.

FRIDAY Data Management

Ms. Richardson's students conducted a survey of their favorite ice cream flavors. Use the pictograph to answer the questions about the results.

Favorite Ice Cream Flavor

Flavor	
Chocolate	
Vanilla	
Strawberry	

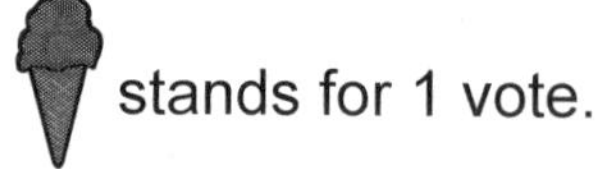

1. How many students liked chocolate? ______________

2. How many students liked vanilla? ________________

3. How many students liked strawberry? ____________

4. How many students voted? ____________________

BRAIN STRETCH

There were 15 ants on a log. 9 ants walked away. How many ants were left?

MONDAY Patterning and Algebra

1. Color the shapes to make a pattern.

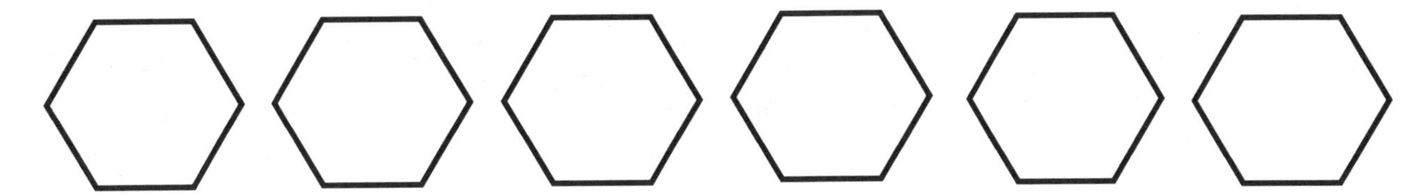

What is your pattern rule? ______________________________

2. Meg was adding 7 + 5 + 3.
I know that 7 + 3 = 10 and
then I can add 5 more.
The answer is 10 + 5 = 15.

Try your own way to add 6 + 8 + 4.

____

3. Count on by 100s from 500.

500, ________, ________, ________

TUESDAY Number Sense and Operations

1. What is the number?

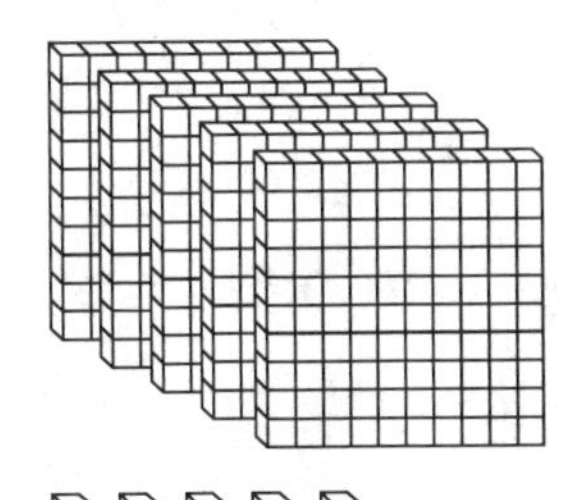

hundreds ______

tens ______

ones ______

number ______

2. A. 574 = ______ hundreds
______ tens
______ ones

B. 682 = ______ hundreds
______ tens
______ ones

3. Circle the third turtle.

4. What is the name of this coin?

A. nickel
B. quarter
C. penny

WEDNESDAY Geometry

1. Circle the name of this shape.

   square triangle

2. How many sides does it have? ________

3. How many vertices does it have? ________

Trace and draw the shape.

THURSDAY Measurement

1. What time is it?

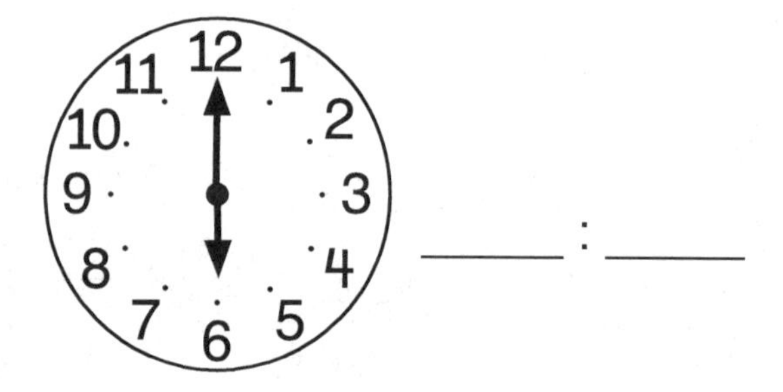

_____ : _____

2. Estimate how long it would take to sneeze.

A. less than one minute
B. more than one minute

3. When do most people have their bedtime?

   A. A.M.

   B. P.M.

4. Measure the length of the line.

It is about ________  long.

FRIDAY Data Management

Here are the results of a Favorite Shape Survey.
Use the data from the pictograph to make a bar graph. Answer the questions.

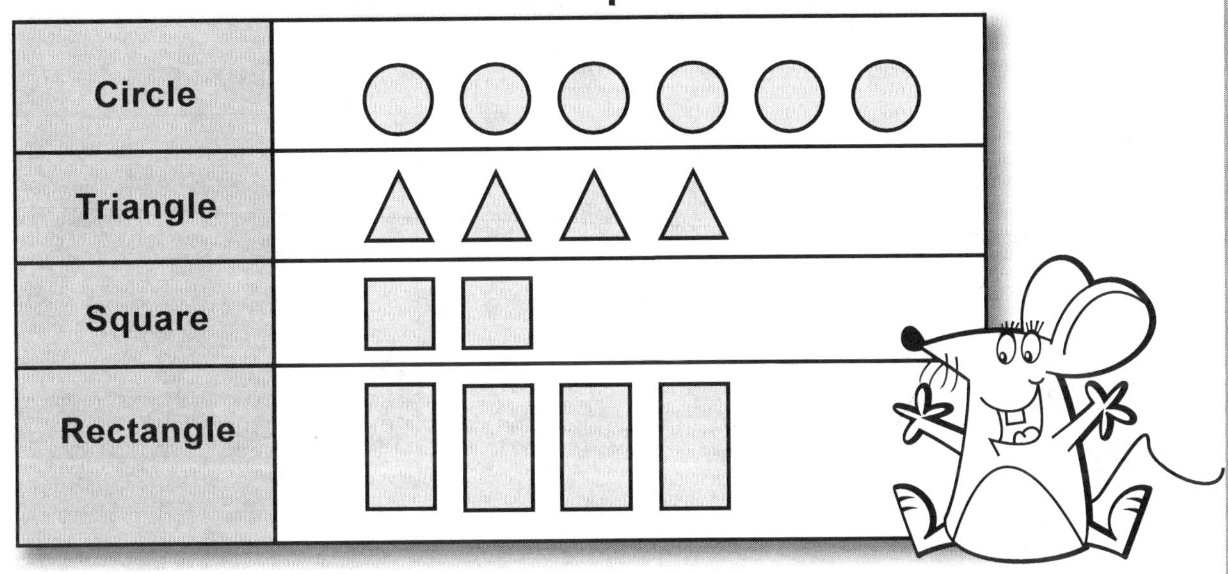

1. How many people answered the survey? ____________________

2. What was the most popular shape? ____________________

3. What was the least popular shape? ____________________

4. Which shapes have the same number of votes? ____________________

BRAIN STRETCH

Carolyn had 18 pieces of bubble gum. She gave 9 pieces to Mike.
How many pieces of bubble gum did she have left?

MONDAY Patterning and Algebra

1. Count by 2s on the chart. Color the numbers.

What patterns do you see?

____

____

1	2	3	4	5	6	7	8	9	10
11	12	13	14	15	16	17	18	19	20
21	22	23	24	25	26	27	28	29	30
31	32	33	34	35	36	37	38	39	40
41	42	43	44	45	46	47	48	49	50
51	52	53	54	55	56	57	58	59	60
61	62	63	64	65	66	67	68	69	70
71	72	73	74	75	76	77	78	79	80
81	82	83	84	85	86	87	88	89	90
91	92	93	94	95	96	97	98	99	100

2. What is the missing number in the sequence?

64, 66, 68, _______, 72, 74,

TUESDAY Number Sense and Operations

1. What is the number?

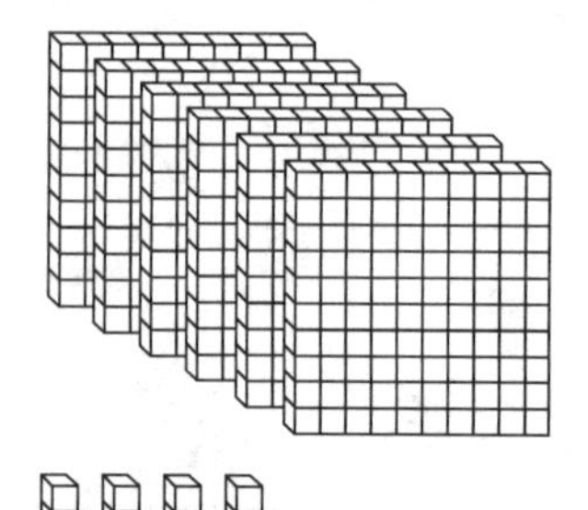

hundreds ______

tens ______

ones ______

number ______

2. Compare the numbers. Use <, >, or =.

A. 58 ☐ 56

B. 99 ☐ 99

3. Circle the first hippopotamus.

4. What is the name of this coin?

A. nickel
B. dime
C. penny

WEDNESDAY Geometry

1. Circle the name of this shape.

   rectangle   triangle

2. How many sides does it have? ________

3. How many vertices does it have? ________

Trace and draw the shape.

THURSDAY Measurement

1. What time is it?

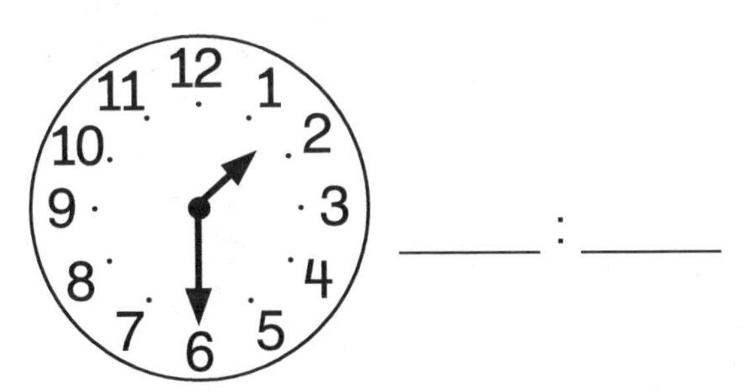

_____ : _____

2. Circle the container that holds more.

A. 1 quart   B. 1 gallon

3. Which is a better estimate for the height of a tree?

   A. 20 feet tall

   B. 20 inches tall

4. Measure the length of the line.

It is about ________  long.

FRIDAY Data Management

Ms. Turnbull's class conducted a survey about favorite kinds of cake. Use the pictograph to answer the questions about the results.

Favorite Cake

Chocolate	🍰 🍰 🍰 🍰 🍰 🍰 🍰 🍰
Vanilla	🍰 🍰 🍰
Strawberry	🍰 🍰 🍰 🍰 🍰 🍰 🍰

Each piece of 🍰 stands for 1 vote.

1. How many students liked chocolate? ______________

2. How many students liked vanilla? ______________

3. How many students liked strawberry? ______________

4. How many students voted? ______________

BRAIN STRETCH

Howard had 22 stamps. He got 10 more.
How many stamps does Howard have in all?

MONDAY Patterning and Algebra

1. Count by 5s on the chart. Color the numbers.

What patterns do you see?

____

____

1	2	3	4	5	6	7	8	9	10
11	12	13	14	15	16	17	18	19	20
21	22	23	24	25	26	27	28	29	30
31	32	33	34	35	36	37	38	39	40
41	42	43	44	45	46	47	48	49	50
51	52	53	54	55	56	57	58	59	60
61	62	63	64	65	66	67	68	69	70
71	72	73	74	75	76	77	78	79	80
81	82	83	84	85	86	87	88	89	90
91	92	93	94	95	96	97	98	99	100

2. What is the missing number in the sequence?

70, 75, 80, ______, 90, 95

TUESDAY Number Sense and Operations

1. What is the number?

tens ______

ones ______

number ______

2. Circle the fourth robot.

3. Order the numbers from least to greatest.

674, 346, 200

________ < ________ < ________

4. Write the number.

A. 10 more than 56 ________

B. 10 less than 72 ________

WEDNESDAY Geometry

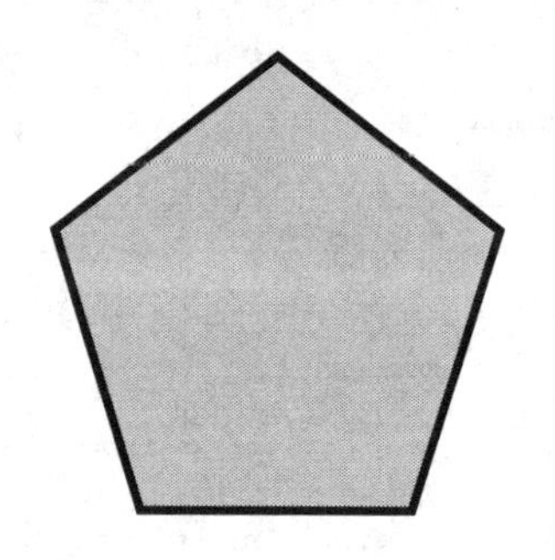

1. Circle the name of this shape.

   rectangle pentagon

2. How many sides does it have? ________

3. How many vertices does it have? ________

Trace and draw the shape.

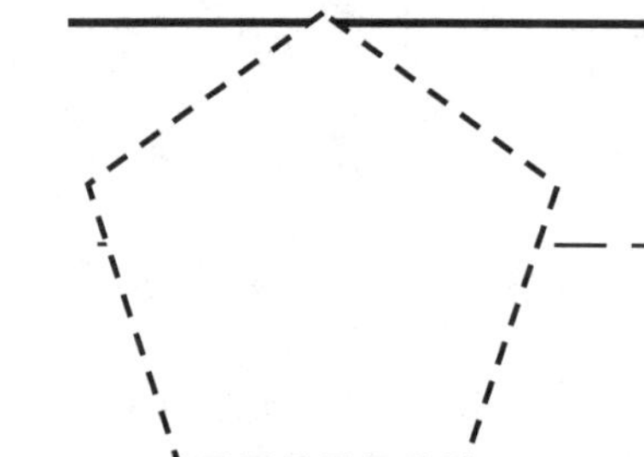

THURSDAY Measurement

1. What time is it?

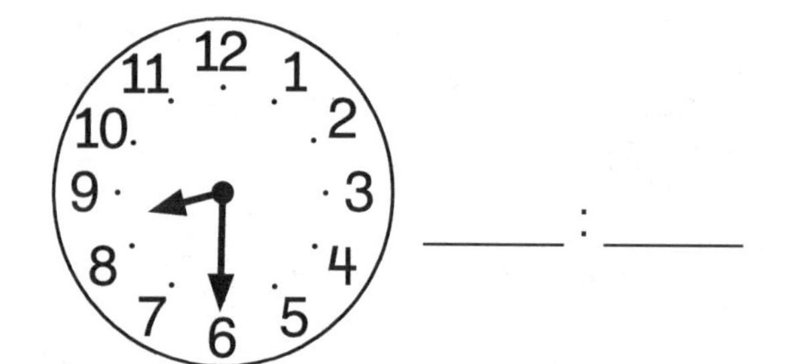

_____ : _____

2. How many inches in 1 foot?

   There are ________ inches in 1 foot.

3. Which tool would be best to tell when it is time for recess?

   A. scale
   B. calendar
   C. clock

4. Measure the length of the line.

It is about ________ long.

FRIDAY Data Management

Ms. Smith's class took a survey of favorite sports.
Use the pictograph to answer the questions about the results.

Favorite Sport

Soccer	☺☺☺☺
Basketball	☺☺☺☺☺☺☺☺
Hockey	☺☺☺☺☺

Each ☺ stands for 2 votes.

1. How many students liked soccer? ____________________

2. How many students liked basketball? ____________________

3. How many students liked hockey? ____________________

4. How many more students liked basketball more than hockey? __________

BRAIN STRETCH

Santos had 13 apples. He needed 20 apples to bake some apple pies.
How many more apples does Santos need?

MONDAY Patterning and Algebra

1. Count by 10s on the chart. Color the numbers.

What patterns do you see?

____

____

2. What is the missing number in the sequence?

40, 50, 60, ______, 80, 90

1	2	3	4	5	6	7	8	9	10
11	12	13	14	15	16	17	18	19	20
21	22	23	24	25	26	27	28	29	30
31	32	33	34	35	36	37	38	39	40
41	42	43	44	45	46	47	48	49	50
51	52	53	54	55	56	57	58	59	60
61	62	63	64	65	66	67	68	69	70
71	72	73	74	75	76	77	78	79	80
81	82	83	84	85	86	87	88	89	90
91	92	93	94	95	96	97	98	99	100

TUESDAY Number Sense and Operations

1. What is the number?

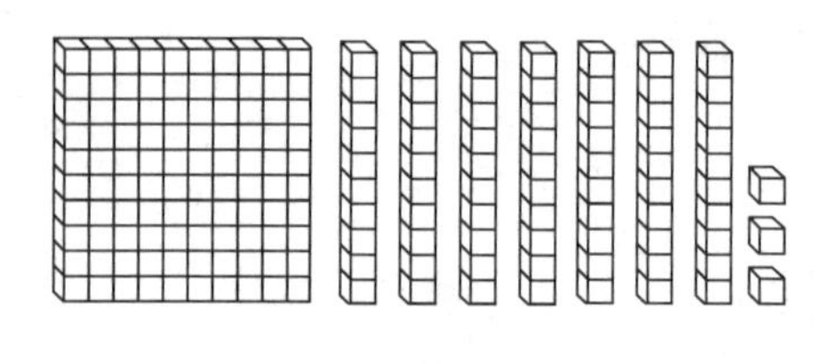

hundreds ______

tens ______

ones ______

number ______

2. Write the number.

A. 10 more than 891 _______

B. 100 less than 211 _______

3. A. 574 = _______ hundreds

_______ tens

_______ ones

B. 682 = _______ hundreds

_______ tens

_______ ones

4. What is the name of this coin?

A. quarter ___________

B. half-dollar _________

C. nickel ____________

WEDNESDAY Geometry

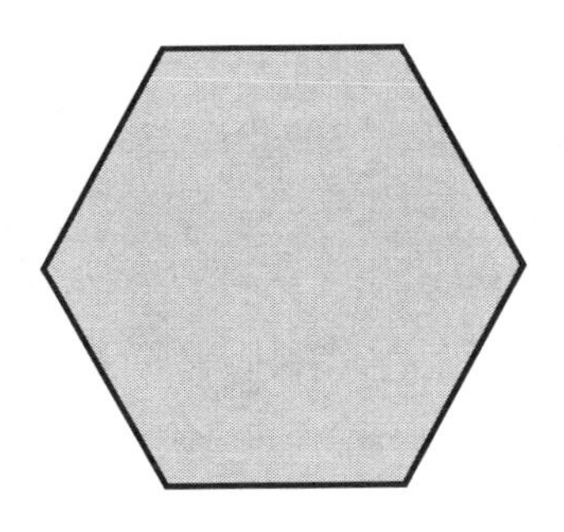

1. Circle the name of this shape.

   rectangle hexagon

2. How many sides does it have? ________

3. How many vertices does it have? ________

Trace and draw the shape.

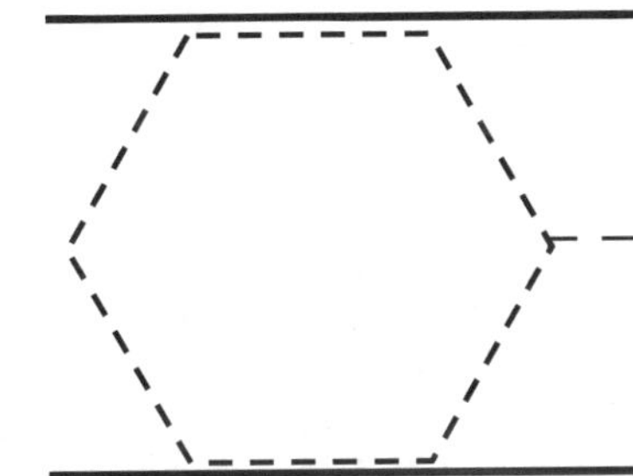

THURSDAY Measurement

1. Which tool would be best to measure the weight of a bag of apples?

   A. scale

   B. ruler

   C. measuring cup

2. What is the better estimate of the length of a car?

   A. more than 1 yard

   B. less than 1 yard

3. Is the temperature hot or cold?

   A. hot

   B. cold

4. Measure the length of the line.

It is about ________  long.

FRIDAY Data Management

Here are the results of a Favorite Zoo Animal Survey.
Use the data from the pictograph to make a bar graph. Answer the questions.

Favorite Zoo Animal

Favorite Zoo Animal

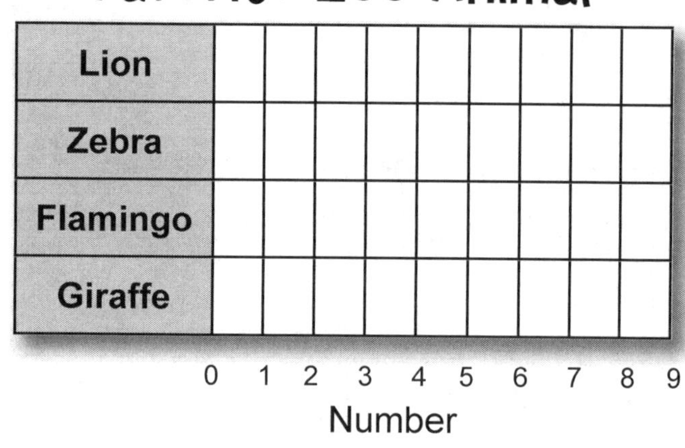

1. How many people answered the survey? ________________

2. What was the most popular zoo animal? ________________

3. What was the least popular zoo animal? ________________

4. How many voted for either a lion or zebra? ________________

BRAIN STRETCH

George had 15 stamps. Carlos had 22 stamps.
How many more stamps did Carlos have?

MONDAY Patterning and Algebra

1. Extend the number pattern.

3, 6, 9, _____, _____, _____

2. There are 7 goldfish in the tank. Maria got 6 more goldfish. How many goldfish are in the tank now? Use pictures and an equation to show your work.

7 + 6 = ☐

_____ goldfish

3. What is the missing number?

_______ − 5 = 8

4. Extend the pattern.

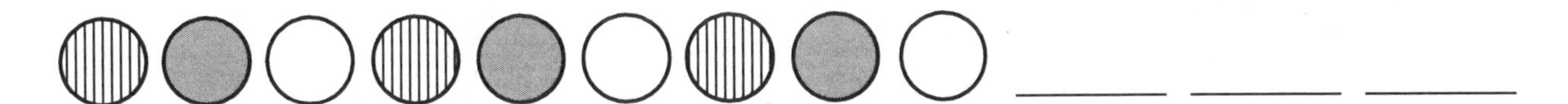

What is your pattern rule? ______________________________

TUESDAY Number Sense and Operations

1. What is the number?

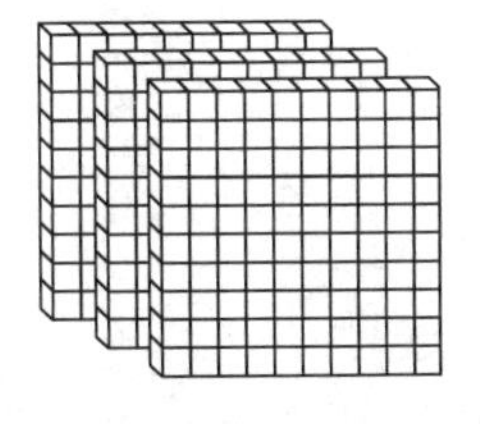

hundreds _____

tens _____

ones _____

number _____

2. Write = or ≠ to make the number sentence true.

15 − 9 ☐ 5 + 4

3. Write the number.

A. 100 more than 452 _______

B. 100 less than 638 _______

4. What is the value of the coins?

_________ ¢

WEDNESDAY Geometry

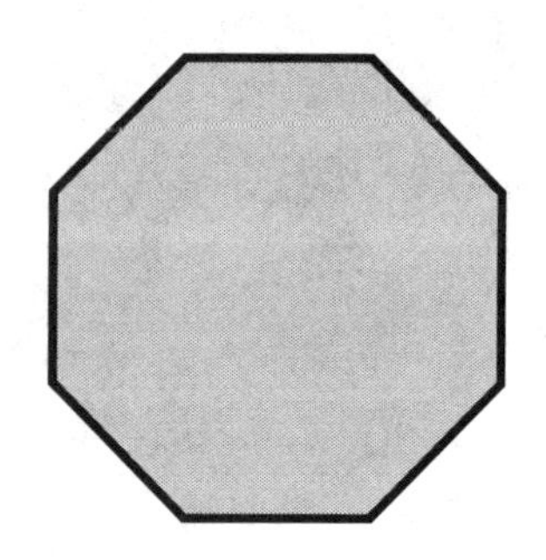

1. Circle the name of this shape.

   triangle   octagon

2. How many sides does it have? ________

3. How many vertices does it have? ________

Trace and draw the shape.

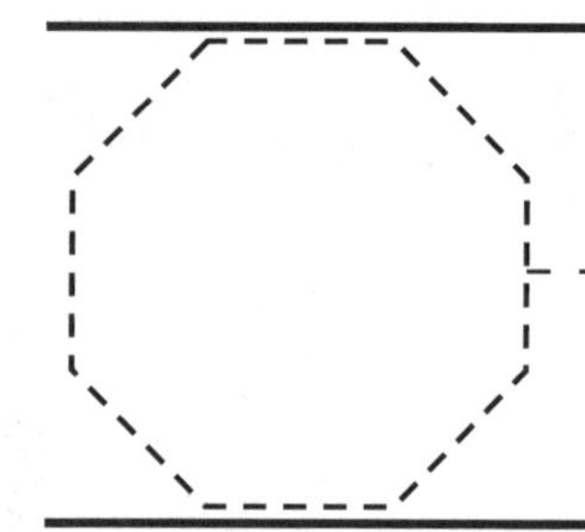

THURSDAY Measurement

1. Draw in the hands on the clock to show the time 5:00.

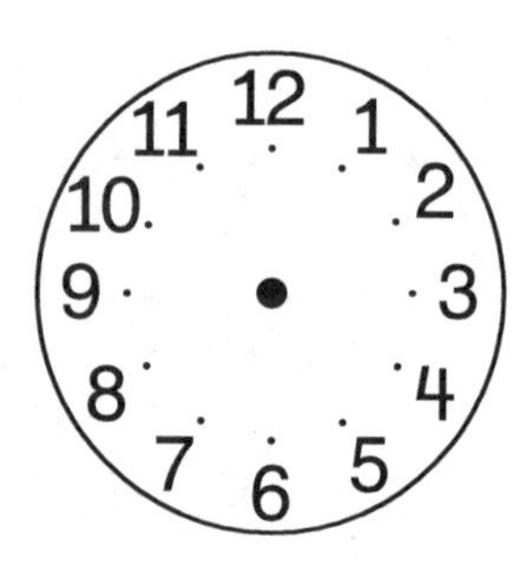

2. Andrew went to the park from 1:00 to 3:00. How long was he at the park?

   ________ hours

3. Draw a line 1 inch long.

4. Choose the better unit of measure for the capacity of a glass of milk.

   A. milliliter

   B. liter

FRIDAY Data Management

Ms. Lopez's class conducted a survey of their favorite fruits.
Use the pictograph to answer the questions about the results.

Favorite Fruit

Orange	
Apple	
Watermelon	

1. How many students liked ? ____________________

2. How many students liked ? ____________________

3. How many students liked ? ____________________

4. How many students voted? ________________________

5. Which fruit was the most popular? ___________________

BRAIN STRETCH

1. $\begin{array}{r} 2 \\ 9 \\ +\,4 \\ \hline \end{array}$

2. $\begin{array}{r} 1 \\ 5 \\ +\,7 \\ \hline \end{array}$

3. $\begin{array}{r} 16 \\ -\;\;7 \\ \hline \end{array}$

4. $\begin{array}{r} 14 \\ -\;\;5 \\ \hline \end{array}$

MONDAY Patterning and Algebra

1. Count on by 5s from 100.

   100, _____, _____, _____, _____

2. There are 9 blue pens and 5 green pens on the table. Tom put 3 more pens on the table. How many pens are there now? Use pictures and the equation to show your work.

   9 + 5 + 3 = ☐

   _____ pens

3. What is the missing number?

   _______ + 7 = 14

4. Extend the pattern.

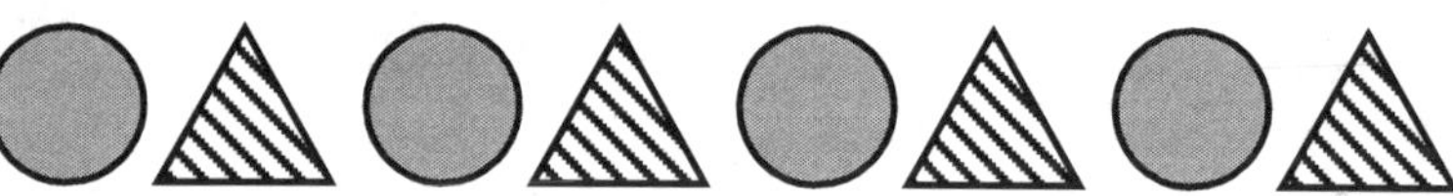
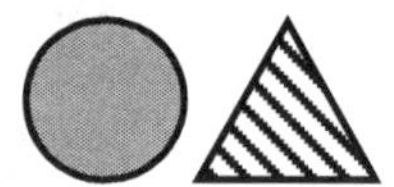

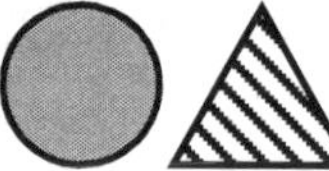

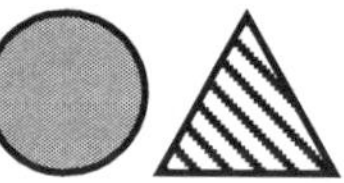

_____ _____ _____

What is your pattern rule? ________________________________

TUESDAY Number Sense and Operations

1. What is the number?

   hundreds _____

   tens _____

   ones _____

   number _____

2. Write = or ≠ to make the number sentence true.

   18 − 10 ☐ 11 + 1

3. What number comes just before?

   A. _______, 81

   B . _______, 36

4. What is the value of the coins?

_________ ¢

WEDNESDAY Geometry

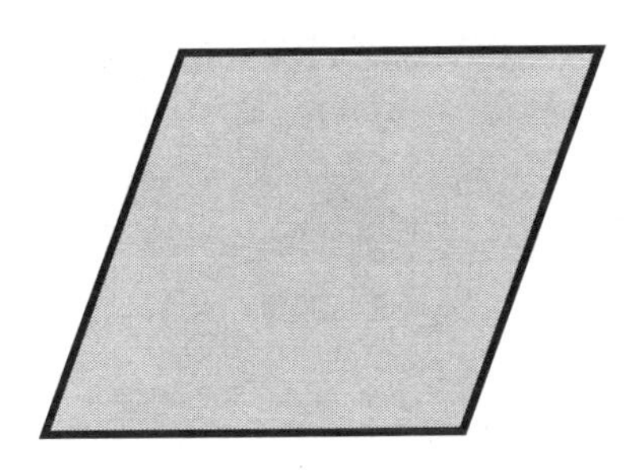

1. Circle the name of this shape.

    parallelogram   circle

2. How many sides does it have? ________

3. How many vertices does it have? ________

Trace and draw the shape.

THURSDAY Measurement

1. What time is it?

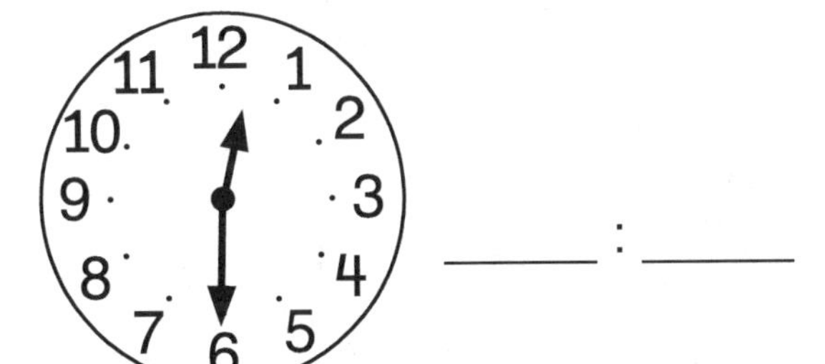

_____ : _____

2. Dave left the house at 3:30 and came back at 7:30. How long was he gone?

    ________ hours

3. What day of the week comes after Monday?

    A. Tuesday
    B. Friday
    C. Sunday

4. Draw a line 3 centimeters long.

FRIDAY Data Management

Here are the results of a Favorite Pet Survey.
Complete the chart and bar graph. Answer the questions about the results.

Favorite Pets Chart

Pet	Tally	Number
Dog		9
Cat		7
Hamster		4
Bird		2

Favorite Pets Graph

Pet											
Dog											
Cat											
Hamster											
Bird											
	0	1	2	3	4	5	6	7	8	9	10	11

Number of People

1. What was the most popular pet? ______________

2. What was the least popular pet? ______________

3. How many people chose either a dog or a bird? ______________

4. How many more people chose a cat than a hamster? ______________

BRAIN STRETCH

1.
$$\begin{array}{r} 3 \\ 8 \\ +\ 2 \\ \hline \end{array}$$

2.
$$\begin{array}{r} 7 \\ 6 \\ +\ 7 \\ \hline \end{array}$$

3.
$$\begin{array}{r} 11 \\ -\ 5 \\ \hline \end{array}$$

4.
$$\begin{array}{r} 18 \\ -\ 9 \\ \hline \end{array}$$

MONDAY Patterning and Algebra

1. What is the missing number in this number pattern?

   55, ______, 75, 85, 95

2. Pat was subtracting 15 – 9.
   I know 9 + 6 = 15
   so 15 – 9 = 6.

   Try Pat's way to solve 13 – 8.

3. Extend the pattern.

     ______ ______

What is your pattern rule? ______________________________

TUESDAY Number Sense and Operations

1. What is the number?

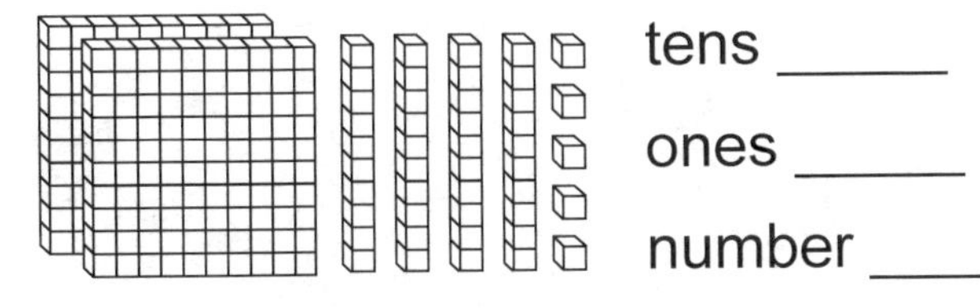

hundreds _____

tens _____

ones _____

number _____

2. Compare the numbers. Use <, >, or =.

   178 ☐ 871

3. Write the number.

   A. 10 more than 325 ________

   B. 10 less than 500 ________

4. What is the value of the coins?

__________ ¢

WEDNESDAY Geometry

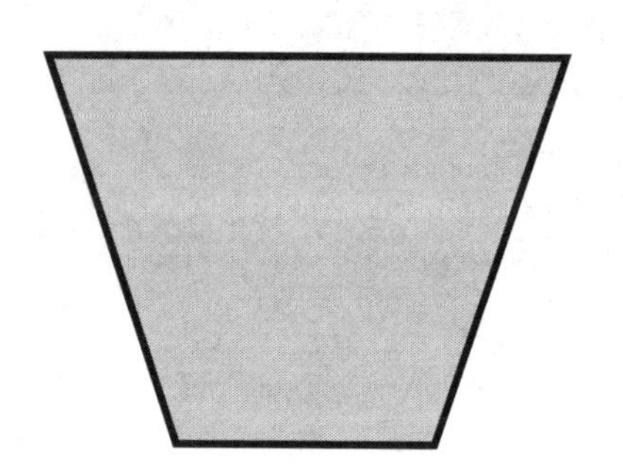

1. Circle the name of this shape.

   trapezoid   circle

2. How many sides does it have? ________

3. How many vertices does it have? ________

Trace and draw the shape.

THURSDAY Measurement

1. What time is it?

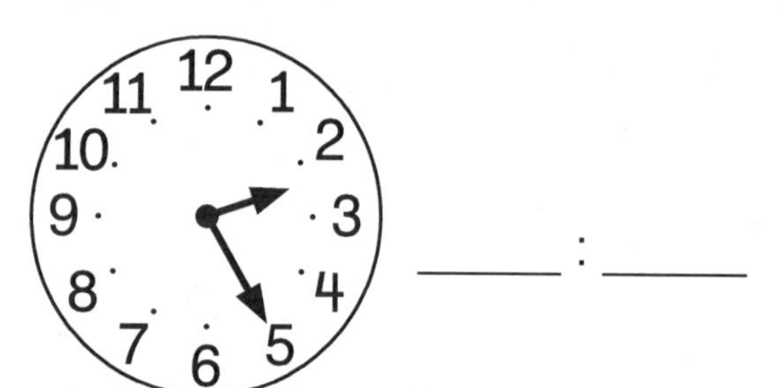

_____ : _____

2. When do you leave for school in the morning?

   A. A.M.

   B. P.M.

3. What day of the week is between Thursday and Saturday?

   ____

4. Which tool would be best to measure the temperature?

   A. scale

   B. thermometer

   C. clock

FRIDAY Data Management

Here are the results of a Favorite Cookie Survey. Complete the chart. Answer the questions about the results.

Favorite Cookie Graph

Chocolate Chip	
Oatmeal Raisin	
Gingerbread	

Each stands for 2 votes.

Favorite Cookie Chart

Cookie	Tally	Number
Chocolate Chip		
Oatmeal Raisin		
Gingerbread		

1. How many students liked chocolate chip? ____________
2. How many students liked oatmeal raisin? ___________
3. How many more students chose chocolate chip than gingerbread? _______
4. Which cookie did students like the most? ______________________________

BRAIN STRETCH

1. $22 + 3 =$ ____
2. $36 - 6 =$ ____
3. $40 + 8 =$ ____
4. $28 - 5 =$ ____

MONDAY Patterning and Algebra

1. What is the missing number in this number pattern?

   32, 42, ______, 62, 72

   Pattern rule ____________________

2. What is the missing number?

   ______ + 4 = 10

3. Write each number in expanded form.

   978 ________________________

   842 ________________________

   731 ________________________

   625 ________________________

4. What is the missing number?

   7 + 9 = 8 + _____

TUESDAY Number Sense and Operations

1. What is the number?

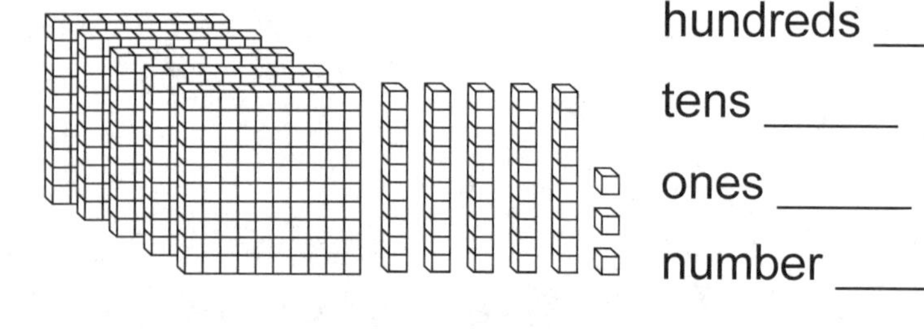

hundreds _____

tens _____

ones _____

number _____

2. What are two related facts for 8 + 2 = 10?

   _____ + _____ = _____

   _____ − _____ = _____

3. Complete the equations.

A. 842 = ______ + ______ + ______

B. 295 = ______ + ______ + ______

4. What is the value of the coins?

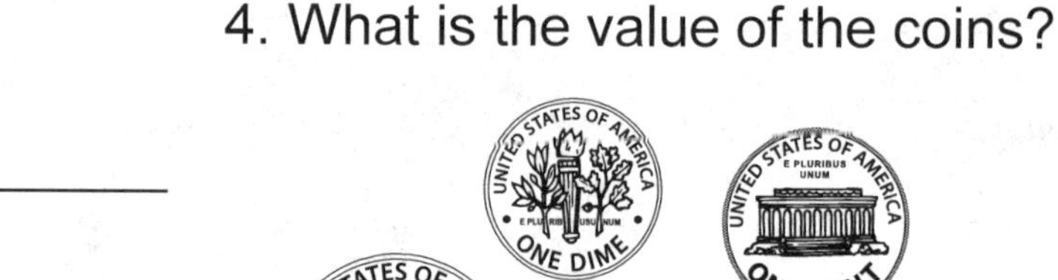

__________ ¢

WEDNESDAY Geometry

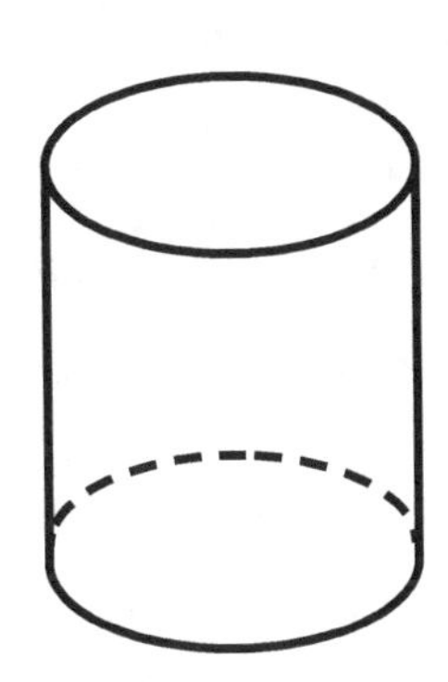

1. Circle the name of this 3D shape.

   cylinder   pyramid

2. How many edges does it have? ________

3. How many faces does it have? ________

4. Look at the shapes. Choose flip, slide, or turn.

A. flip
B. slide
C. turn

THURSDAY Measurement

1. What time is it?

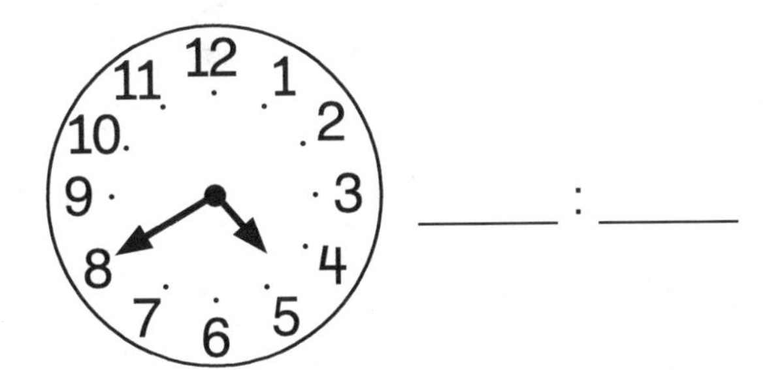

_____ : _____

2. What time will it be in 4 hours?

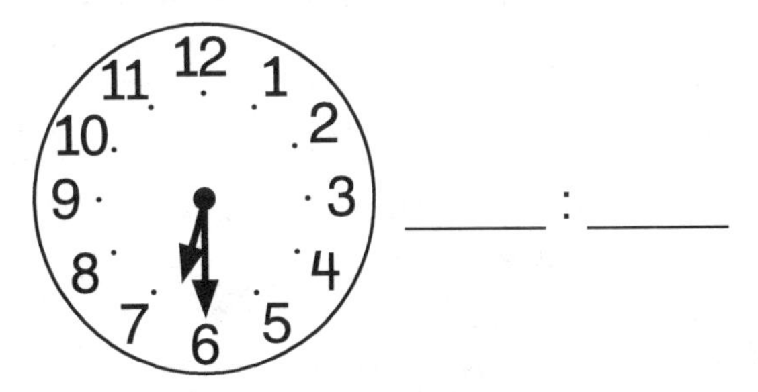

_____ : _____

3. How many months in a year?

__________ months

4. Which tool would be best to measure the width of a window?

A. scale
B. yardstick
C. clock

FRIDAY Data Management

Count the pictures and complete the favorite shape bar graph. Answer the questions.

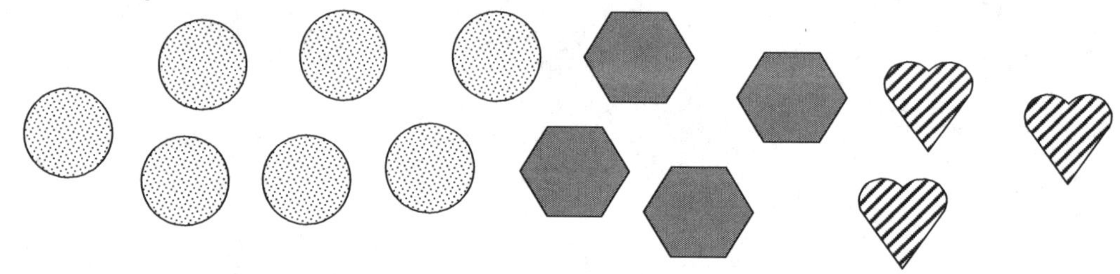

Favorite Shape Graph

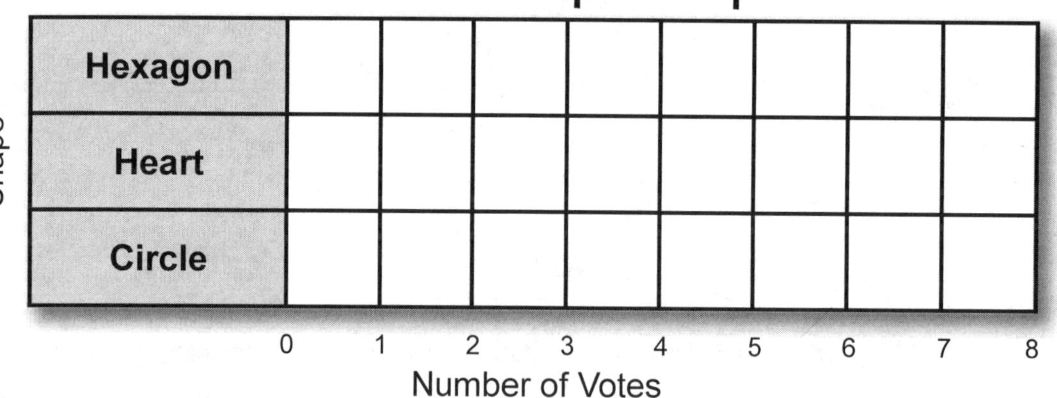

Shape								
Hexagon								
Heart								
Circle								

0 1 2 3 4 5 6 7 8

Number of Votes

1. Which shape is the most popular? ________________

2. Which shape is the least popular? ________________

3. How many votes altogether? ________________

4. How many more votes for circle than heart? ________________

BRAIN STRETCH

1. $\begin{array}{r} 15 \\ +\ 2 \\ \hline \end{array}$

2. $\begin{array}{r} 20 \\ -\ 6 \\ \hline \end{array}$

3. $\begin{array}{r} 23 \\ +\ 4 \\ \hline \end{array}$

4. $\begin{array}{r} 38 \\ -\ 7 \\ \hline \end{array}$

MONDAY Patterning and Algebra

1. What is the missing number?

_______ + 5 = 11

2. There are 4 balls in a box. Jill put some more balls in the box. There are now 13 balls in the box. How many balls did Jill put in? Use pictures and the equation to show your work.

4 + ☐ = 13

_____ balls

3. Extend the pattern.

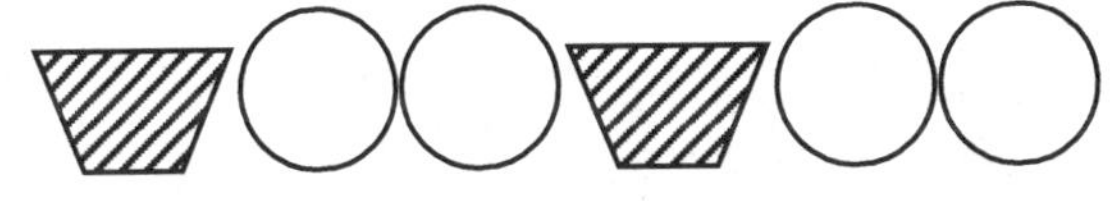 _______, _______, _______, _______

What is the pattern rule? ______________________________

TUESDAY Number Sense and Operations

1. What is the number?

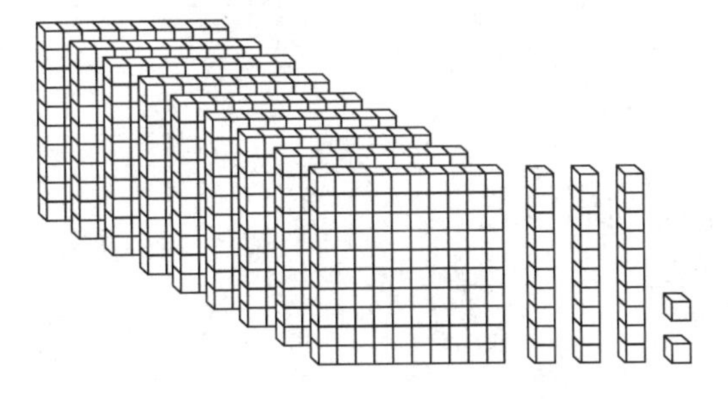

hundreds _____

tens _____

ones _____

number _____

2. Pair the circles. Are any circles left over? _____

Is the number 4 odd or even? _____

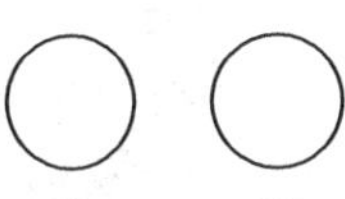

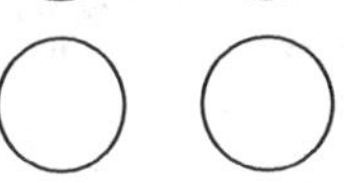

3. Color to show $\frac{3}{4}$.

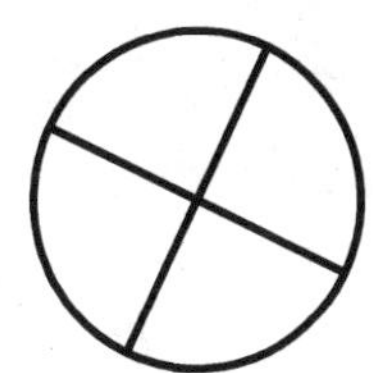

4. What is the value of the coins?

__________ ¢

WEDNESDAY Geometry

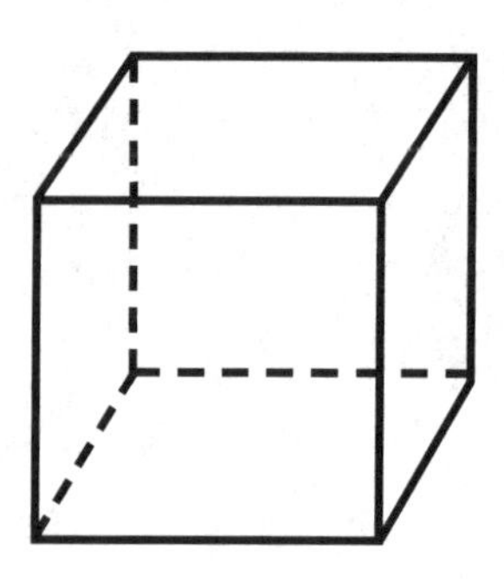

1. Circle the name of this 3D shape.

   cube pyramid

2. How many edges does it have? ________

3. How many faces does it have? ________

4. Draw a circle.

5. What shape is inside the square?

A. triangle
B. square

THURSDAY Measurement

1. What time is it?

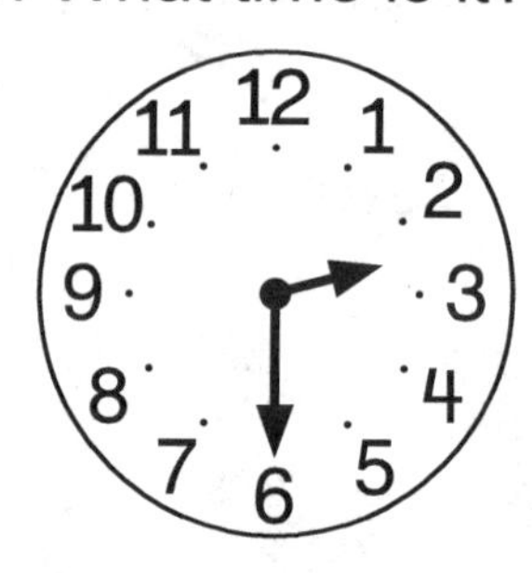

half past ________

2. What day of the week comes just before Wednesday?

   A. Tuesday
   B. Friday
   C. Monday

3. When do most people eat dinner?

   A. A.M.
   B. P.M.

4. What is a better estimate for the length of a classroom?

   A. 30 feet
   B. 30 inches

FRIDAY Data Management

Here are the results of a favorite vegetable survey.
Use the data from the bar graph to answer the questions.

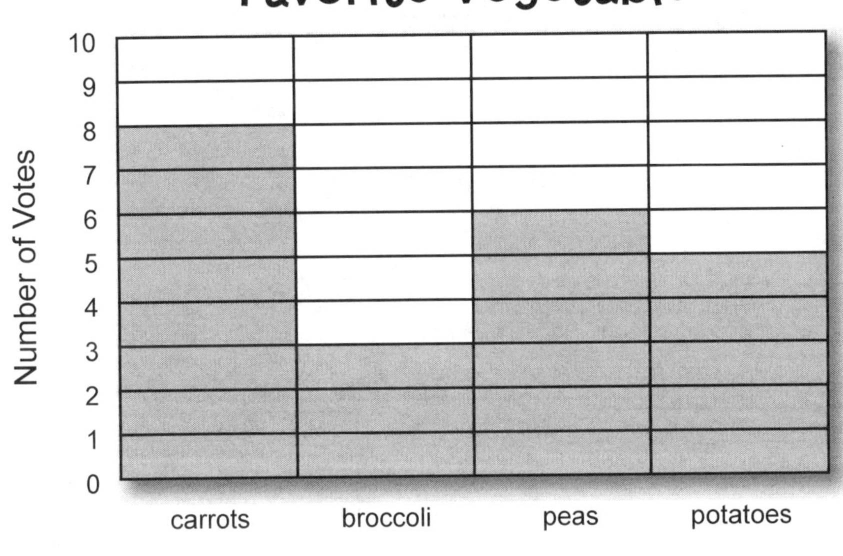

1. Which vegetable has the least votes? ____________________

2. How many more votes are there for carrots than broccoli? ____________

3. Which vegetable is the most popular? ____________________

4. How many votes for peas and potatoes combined? ____________

BRAIN STRETCH

1. 79 + 15

2. 31 + 29

3. 98 − 68

4. 87 − 55

MONDAY Patterning and Algebra

1. What is the missing number?

   _______ + 8 = 11

2. There are 7 soccer balls in the gym. Lewis took 4 balls outside. Clark put 8 more soccer balls in the gym. How many soccer balls are in the gym now? Use pictures and the equation to show your work.

   7 – 4 + 8 = ☐

   ____ soccer balls

3. Count on by 10s from 920.

   920, ______, ______, ______, ______

4. What is the missing number to complete the number sentence?

   ______ + 9 = 2 + 7

TUESDAY Number Sense and Operations

1. What is the number?

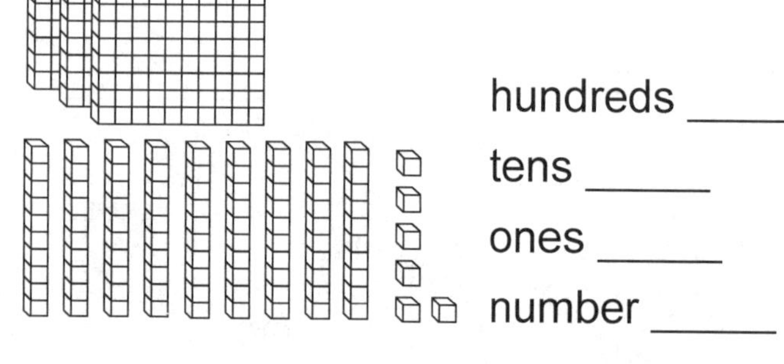

   hundreds ______
   tens ______
   ones ______
   number ______

2. Compare the numbers. Use <, >, or =.

   341 ☐ 351

3. Write the number.

   A. 100 less than 551 ________

   B. 100 more than 280 ________

4. What is the value of the coins?

   ________ ¢

WEDNESDAY Geometry

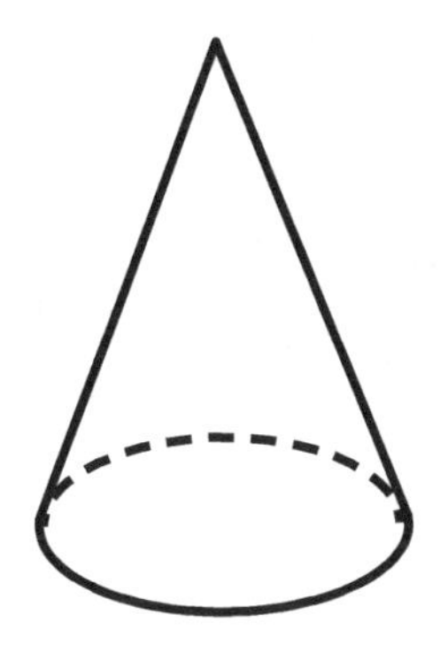

1. Circle the name of this 3D shape.

   cone   pyramid

2. How many edges does it have? ________

3. How many faces does it have? ________

4. Draw a triangle.

5. What shape is inside the square?

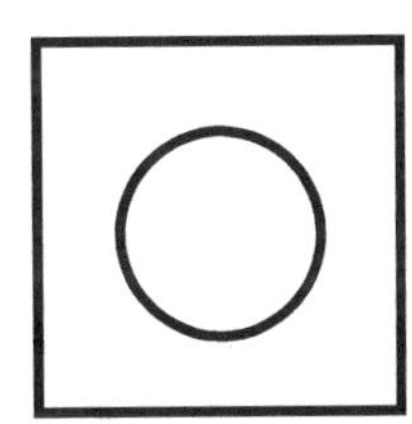

A. circle
B. square

THURSDAY Measurement

1. Write the time in two ways.

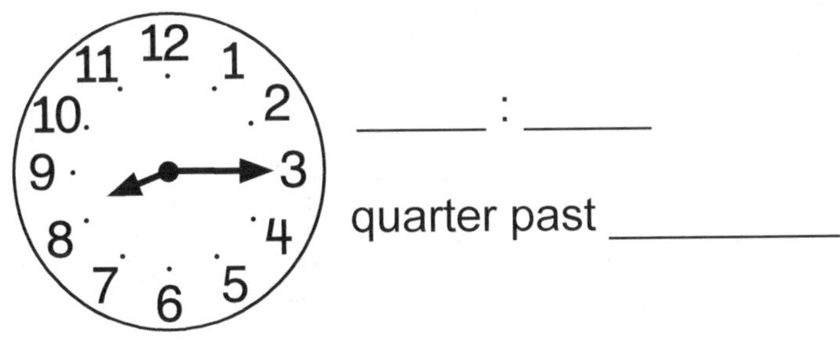

_____ : _____

quarter past __________

2. What is the better estimate of the weight of a crayon?

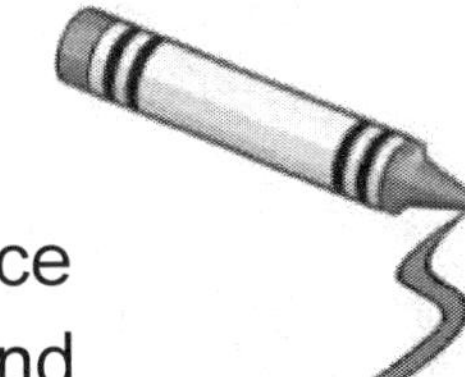

A. 1 ounce
B. 1 pound

3. What day of the week is just after Thursday?

   A. Monday
   B. Tuesday
   C. Friday

4. Measure the length of the line.

It is about __________ [paper clip] long.

FRIDAY Data Management

Use the calendar to answer the questions.

June

Sunday	Monday	Tuesday	Wednesday	Thursday	Friday	Saturday
			1	2	3	4
5	6	7	8	9	10	11
12	13	14	15	16	17	18
18	20	21	22	23	24	25
26	27	28	29	30		

1. How many days are there in the month of June? ____________________

2. What day of the week is June 20th? ____________________

3. How many Thursdays are in June? ____________________

4. What day of the week will July start on? ____________________

BRAIN STRETCH

1. $\begin{array}{r} 26 \\ +\ 35 \\ \hline \end{array}$

2. $\begin{array}{r} 74 \\ +\ 36 \\ \hline \end{array}$

3. $\begin{array}{r} 43 \\ -\ 18 \\ \hline \end{array}$

4. $\begin{array}{r} 92 \\ -\ 35 \\ \hline \end{array}$

MONDAY Patterning and Algebra

1. What is the missing number to complete the number sentence?

   9 − 6 = 2 + _____

2. How many circles are there in all? Add the rows. Write the addition equation.

   ___ + ___ + ___ + ___ = ___

   ○○○
   ○○○
   ○○○
   ○○○

3. Make a pattern using ○ and □.

   ____

   Explain your pattern rule. ____

TUESDAY Number Sense and Operations

1. What is the number?

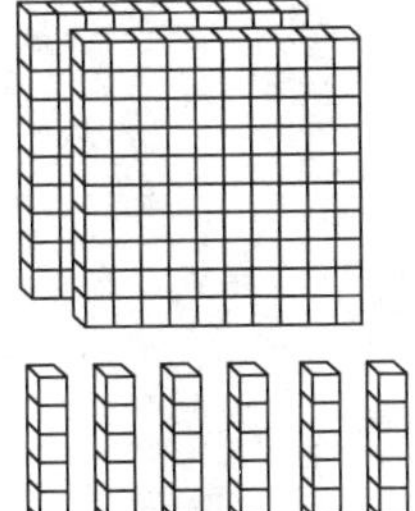

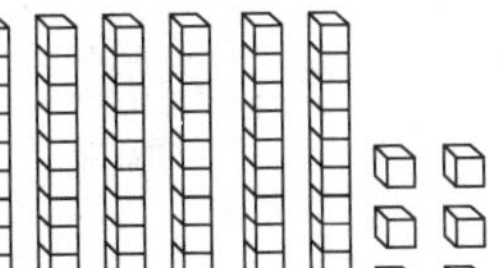

hundreds _____

tens _____

ones _____

number _____

2. Compare 287 and 289. Fill in the table.

	Hundreds	Tens	Ones
287			
289			

The hundreds are the same. The tens are the same. But 287 has _____ less ones than 289.

Use < or >. 287 □ 289

3. Write the numbers.

   A. two hundred sixty-three _______

   B. 500 + 60 + 9 = _______

4. What is the value of the coins?

_______ ¢

WEDNESDAY Geometry

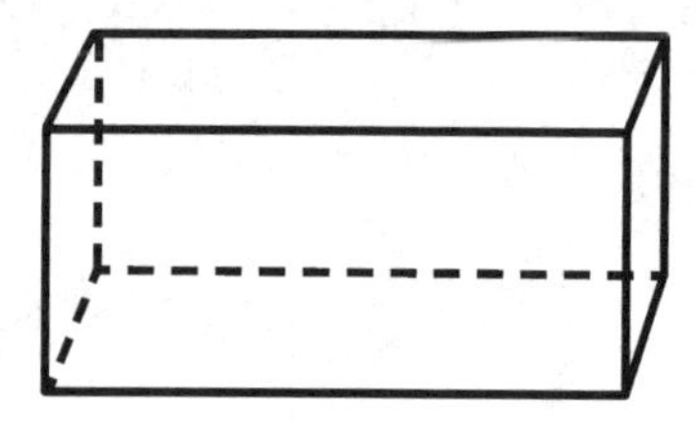

1. Circle the name of this 3D shape.

   rectangular prism   pyramid

2. How many edges does it have? ________

3. How many faces does it have? ________

4. Draw a square.

5. What shape is inside the triangle?

A. triangle
B. square

THURSDAY Measurement

1. Write the time in two ways.

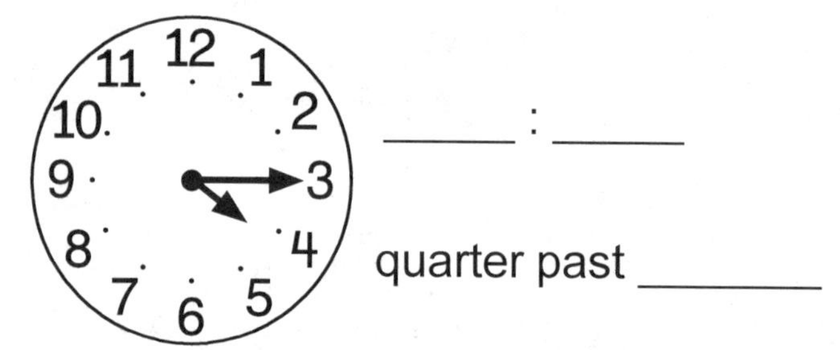

2. About how wide is your finger?

   A. 1 meter
   B. 1 centimeter

3. What day of the week is just before Sunday?

   A. Saturday
   B. Tuesday
   C. Friday

4. What is a better estimate for the length of a truck?

   A. 16 feet
   B. 16 inches

FRIDAY Data Management

Here are the results of a Favorite Season Survey.
Use the data from the pictograph to make a bar graph. Answer the questions.

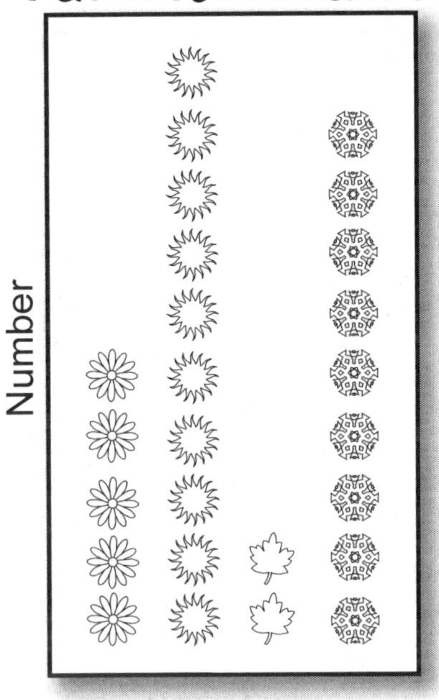

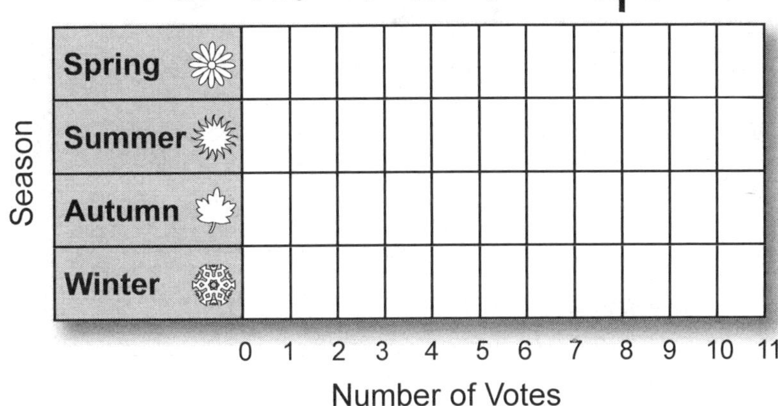

1. How many people voted for spring and summer? ____________________

2. How many more people voted for winter than spring? ________________

3. Which season had the least number of votes? ____________________

4. How many people voted for winter and summer? ____________________

BRAIN STRETCH

1. $\begin{array}{r} 45 \\ +\ 17 \\ \hline \end{array}$

2. $\begin{array}{r} 69 \\ +\ 14 \\ \hline \end{array}$

3. $\begin{array}{r} 87 \\ -\ 28 \\ \hline \end{array}$

4. $\begin{array}{r} 70 \\ -\ 39 \\ \hline \end{array}$

MONDAY Patterning and Algebra

1. What is the missing sign?

3 _______ 12 = 15

2. What is the next number if the pattern rule is add 5?

11, _____

3. Make an AB pattern using  .

____

4. Count on by 2s from 88.

88, ________, ________, ________, ________

TUESDAY Number Sense and Operations

1. Draw a model for 363 using

□ hundred | ten • one

2. Is 5 an odd or an even number?

______

Pair the circles and explain your thinking.

____

○○○○○

3. Compare 562 and 566.
Fill in the table.

	Hundreds	Tens	Ones
562			
566			

The hundreds are ______.

The tens are ______.

562 has _____ less ones than 566.

Use < or >. 562 □ 566

4. What is the value of the coins?

________ ¢

WEDNESDAY Geometry

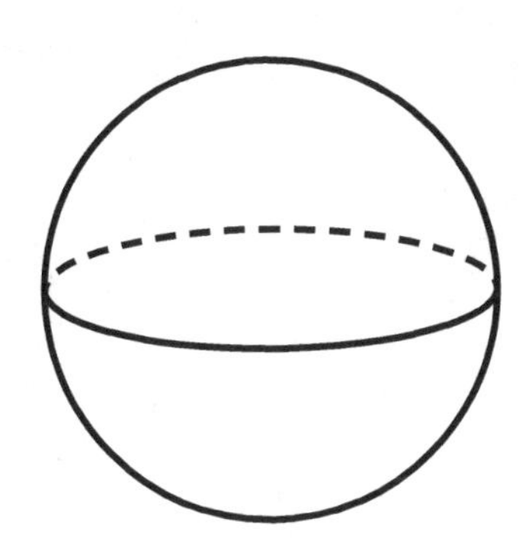

1. Circle the name of this 3D shape.

cylinder sphere

2. How many edges does it have? ________

3. How many faces does it have? ________

4. Draw a rectangle.

5. What shape is beside the square?

A. triangle
B. square

THURSDAY Measurement

1. Write the time in two ways.

12 1 2 3 4 5 6 7 8 9 10 11

_____ : _____

quarter past ________

2. How many feet in 1 yard?

There are ________ feet in 1 yard.

3. How many weeks are there in a year?

________ weeks

4. Measure the length of the line.

It is about ___________ long.

FRIDAY Data Management

Here are the results of a Favorite Breakfast Food Survey.
Complete the chart and answer the questions about the results.

Favorite Breakfast Food

Favourite Breakfast Foods	Tally	Number
Cereal		5
Eggs		10
Pancakes		6
Granola		6

1. What was the most popular breakfast food? ____________________
2. How many people liked either cereal or pancakes? ____________________
3. Which breakfast foods did people like the same? ____________________
4. What was the least popular breakfast food? ____________________

BRAIN STRETCH

1. $\begin{array}{r} 18 \\ +\ 76 \\ \hline \end{array}$

2. $\begin{array}{r} 39 \\ +\ 47 \\ \hline \end{array}$

3. $\begin{array}{r} 90 \\ -\ 27 \\ \hline \end{array}$

4. $\begin{array}{r} 81 \\ -\ 39 \\ \hline \end{array}$

MONDAY Patterning and Algebra

1. Write = or ≠ to make the number sentence true.

   15 − 9 ☐ 5 + 7

2. What is the next number if the pattern rule is subtract 2?

   10, ____

3. What is the missing number to make the equation true?

   2 + 7 + 3 = ________ + 6 + 1

4. There were 7 snakes in the garden. Some more snakes came. Now there are 19 snakes. How many snakes came to the garden? Use pictures and an equation to show your work.

   ____

   ____ snakes

TUESDAY Number Sense and Operations

1. Draw a model for 479 using

   ☐ hundred | ten • one

2 Circle the value of the underlined digit.

A.	6<u>7</u>1	700	70	7
B.	<u>3</u>58	300	30	3

3. Is the number odd or even? Count by 2s to find out.

   A. 14 ______

   B. 5 ______

4. What is the value of the coins?

______ ¢

WEDNESDAY Geometry

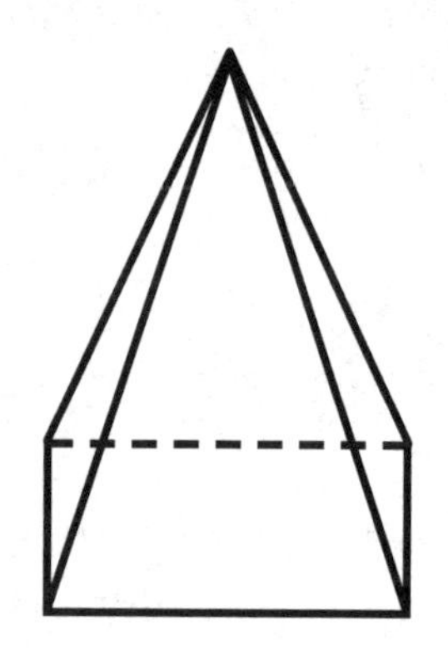

1. Circle the name of this 3D shape.

   cylinder   pyramid

2. How many edges does it have? ________

3. How many faces does it have? ________

4. Draw a pentagon.

5. What shape is under the circle?

A. triangle
B. square

THURSDAY Measurement

1. Compare the lengths.
   Which length is equal to the line?

A. 🦄🦄🦄🦄🦄🦄🦄

B. 🦄🦄🦄🦄🦄🦄🦄🦄🦄🦄🦄

2. There were 37 students in the gym. 14 more students came into the gym. How many students are in the gym? Use the number line to help solve.

   ____ students

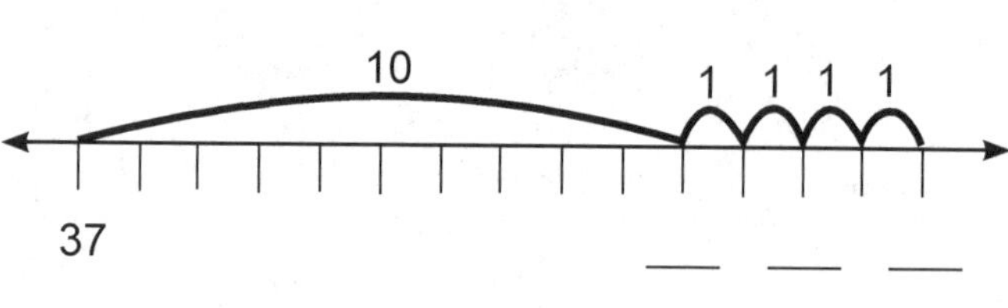

3. How many days are there in a year?

   ________ days

4. Draw a line 5 centimeters long.

FRIDAY Data Management

Here are the results of a Favorite Drink Survey.
Use the pictograph to answer the questions about the results.

Favorite Drink

Lemonade	
Milk	
Water	
Juice	

Each [glass] stands for 2 votes.

1. What drink do people like the most? ______________________

2. What drink do people like the least? ______________________

3. How many people voted in this survey? ______________________

4. Which two drinks did people like the same? ______________________

BRAIN STRETCH

1. $68 + 14 =$ ____

2. $72 + 19 =$ ____

3. $52 - 37 =$ ____

4. $41 - 39 =$ ____

MONDAY Patterning and Algebra

1. Write = or ≠ to make the number sentence true.

   5 + 12 ☐ 10 + 10

2. What is the next number if the pattern rule is add 3?

   4, ________

3. What is the missing number to make the equation true?

   1 + 7 + ______ = 4 + 4 + 4

4. How many circles are there in all? Add the columns. Write the equation.

   ___ + ___ + ___ + ___ = ___

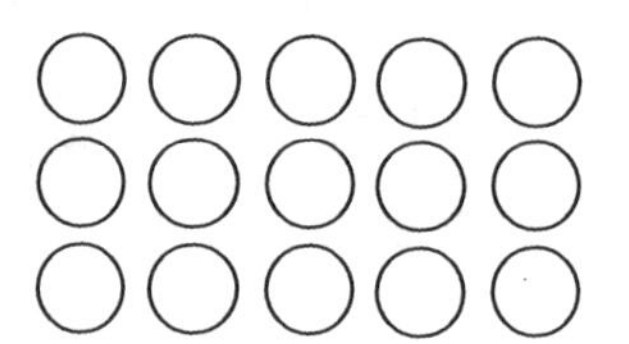

TUESDAY Number Sense and Operations

1. What is the number?

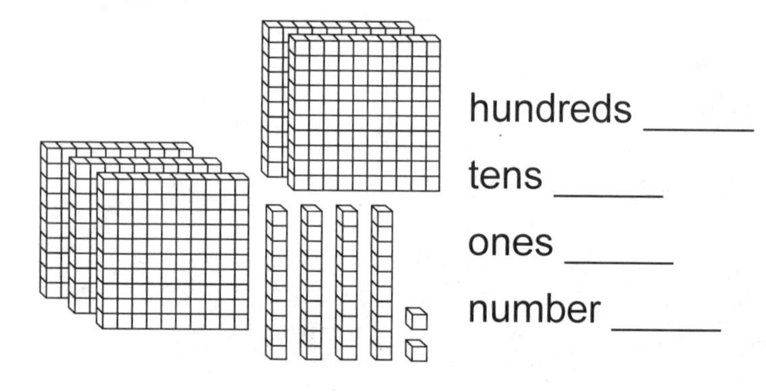

   hundreds _____

   tens _____

   ones _____

   number _____

2. Order the numbers from the greatest to least.

   73, 21, 17

   ______ > ______ > ______

3. What are two related facts for 5 + 3 = 8?

   _____ + _____ = _____

   _____ − _____ = _____

4. Color $\frac{1}{2}$ of the shape.

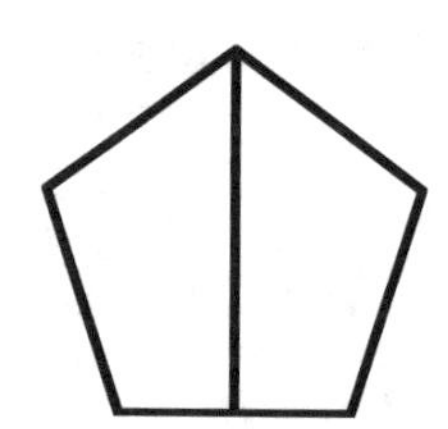

WEDNESDAY Geometry

1. Color the shapes that are the same size and shape.

2. Draw a hexagon.

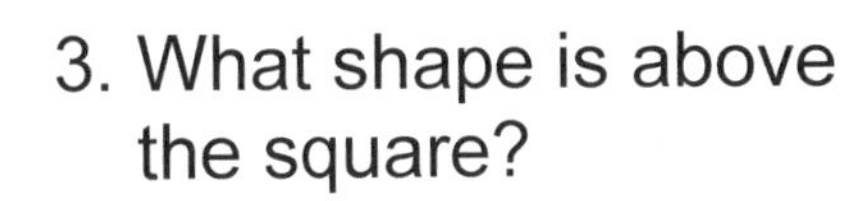

3. What shape is above the square?

A. triangle

B. square

THURSDAY Measurement

1. Write the time in two ways.

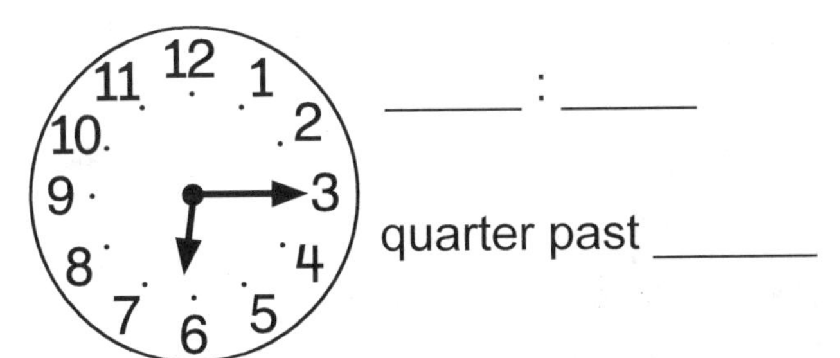

_____ : _____

quarter past _____

2. Circle the better estimate of the length of a strawberry.

A. 1 inch

B. 1 foot

3. Is the temperature hot or cold?

A. hot

B. cold

4. Measure the length of the line.

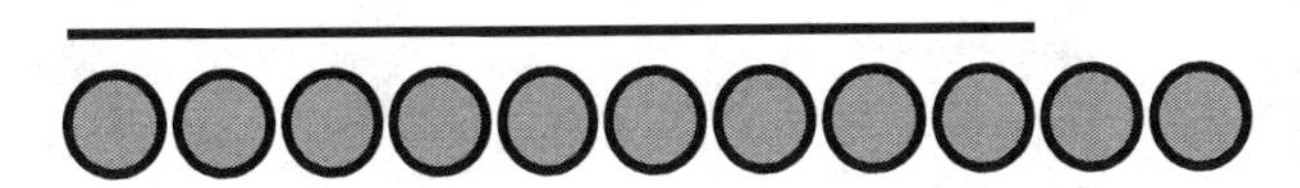

It is about __________ ● long.

FRIDAY Data Management

Use the Venn diagram to answer the questions about these students' favorite recess activities.

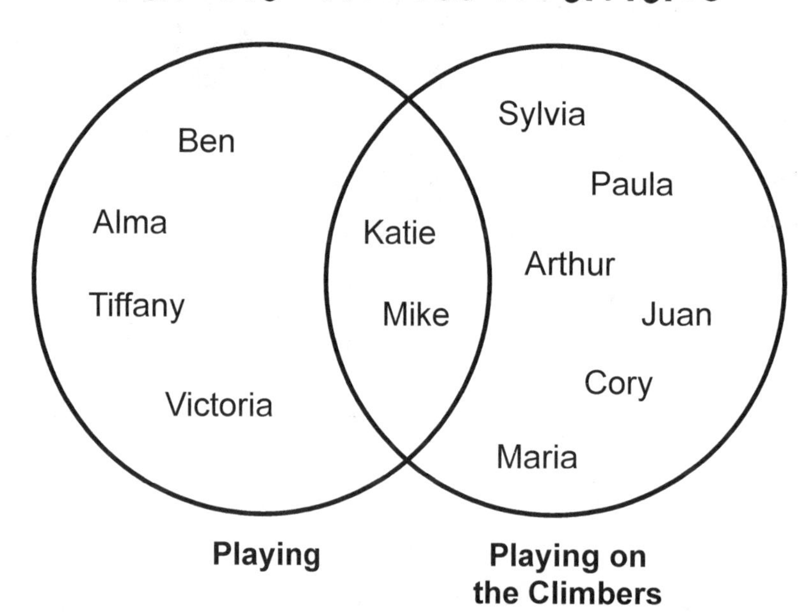

1. Which students like to play tag, but not play on the climbers?

____

2. Which students like to play on the climbers, but not play tag?

____

3. Which students like to do both activities?

____

BRAIN STRETCH

1. $\begin{array}{r} 45 \\ +\ 39 \\ \hline \end{array}$

2. $\begin{array}{r} 61 \\ +\ 29 \\ \hline \end{array}$

3. $\begin{array}{r} 70 \\ -\ 44 \\ \hline \end{array}$

4. $\begin{array}{r} 35 \\ -\ 19 \\ \hline \end{array}$

MONDAY Patterning and Algebra

1. Write = or ≠ to make the number sentence true.

   14 − 7 ☐ 12 − 5

2. What is the next number if the pattern rule is subtract 7?

   14, ________

3. What is the missing number to make the equation true?

   2 + 5 + ________ = 3 + 3 + 3

4. What is the missing number in this sequence?

   ________, 630, 640, 650, 660

TUESDAY Number Sense and Operations

1. Draw a model for 295 using

   ☐ hundred | ten • one

2. What are two related facts for 7 + 5 = 12?

   ______ + ______ = ______

   ______ − ______ = ______

3. Write the number.

   A. 800 + 10 + 2 = ________

   B. 500 + 40 + 1 = ________

4. Color $\frac{1}{3}$ of the shape.

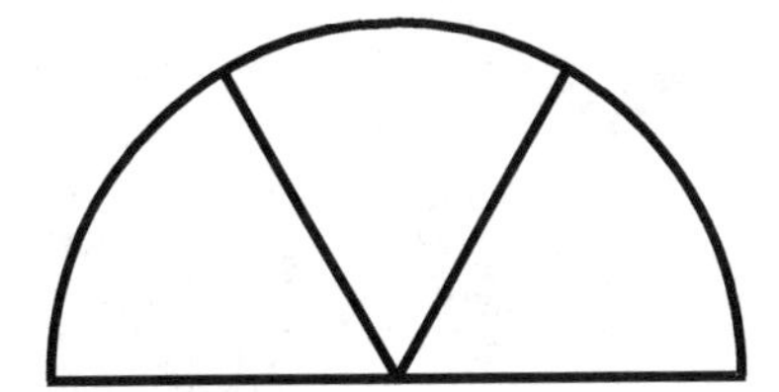

WEDNESDAY Geometry

1. Color the shapes that are the same size and shape.

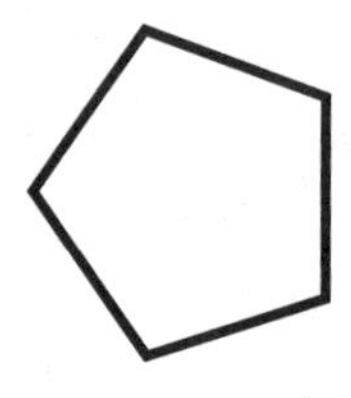 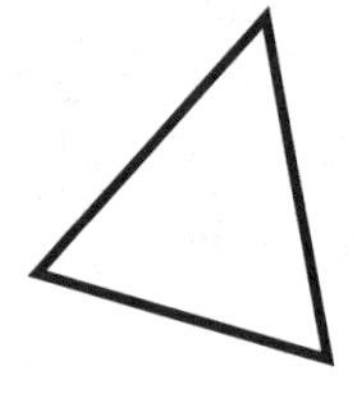 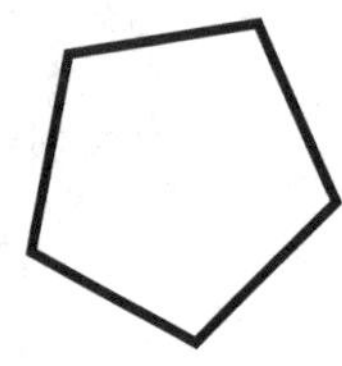  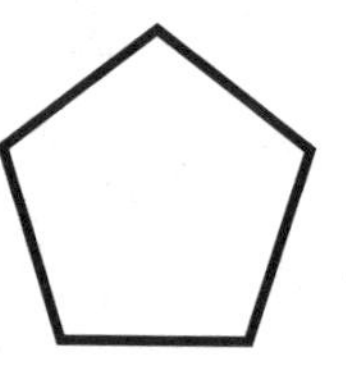 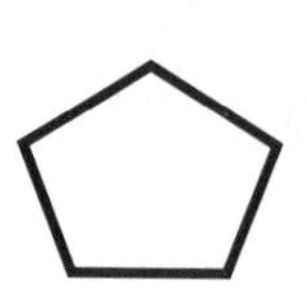

2. Draw an octagon.

3. What shape is inside the triangle?

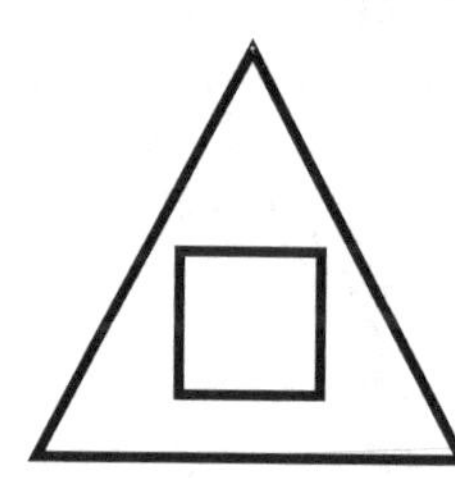

A. circle

B. square

THURSDAY Measurement

1. Write the time in two ways.

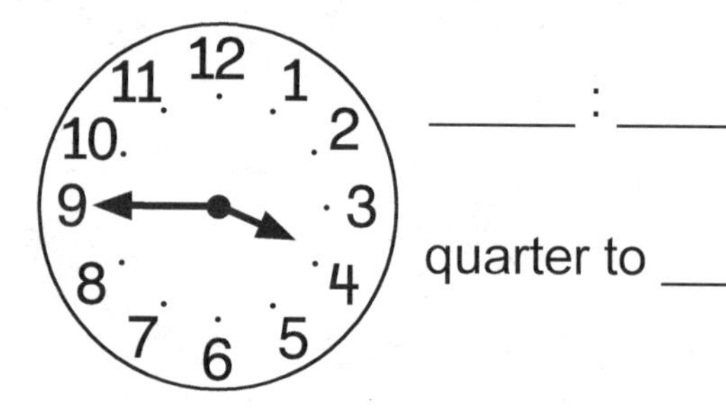

_____ : _____

quarter to _______

2. Compare the lengths.
Which length is longer than the line?

A. 

B. 

3. Which tree is taller?

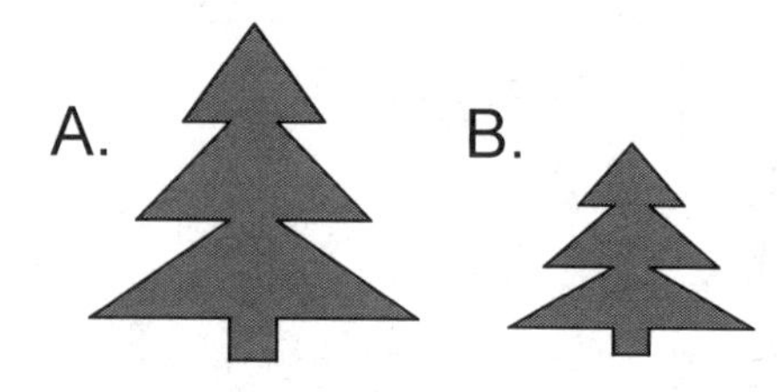

4. Measure the length of the line.

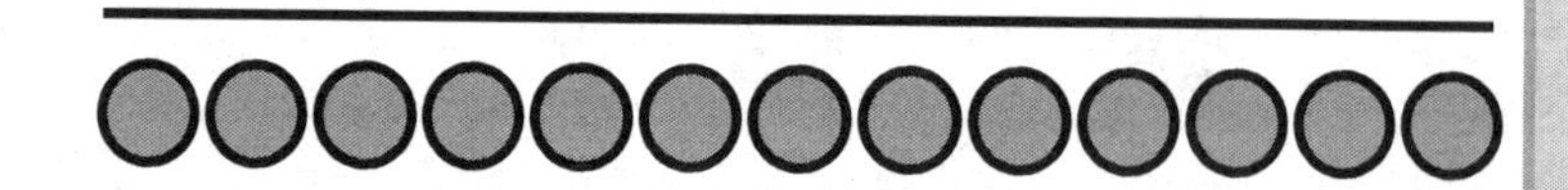

It is about _________  long.

FRIDAY Data Management

Use the Venn diagram to answer the questions about these students' favorite snack foods.

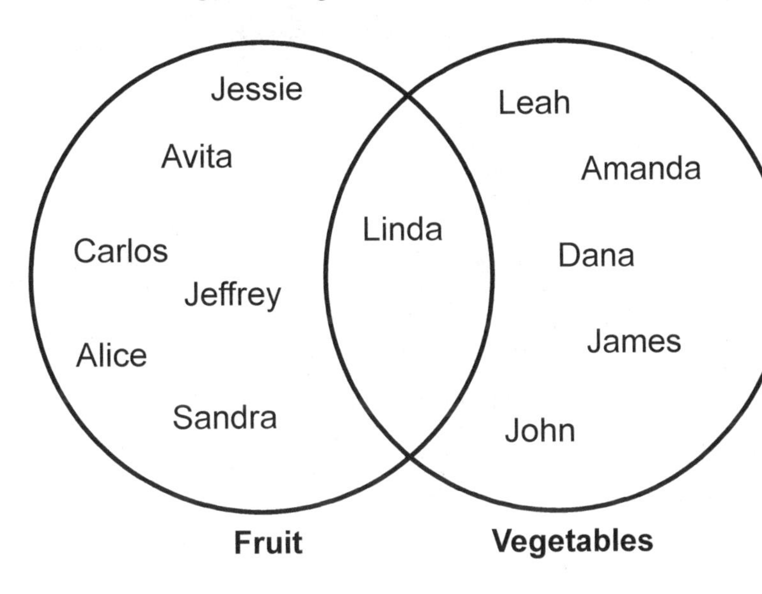

1. Which students like fruit, but not vegetables?

____

2. Which students like to eat vegetables as a snack?

____

3. Which students like vegetables, but not fruit?

____

BRAIN STRETCH

1. $\begin{array}{r} 37 \\ +\ 58 \\ \hline \end{array}$

2. $\begin{array}{r} 29 \\ +\ 46 \\ \hline \end{array}$

3. $\begin{array}{r} 65 \\ -\ 37 \\ \hline \end{array}$

4. $\begin{array}{r} 58 \\ -\ 29 \\ \hline \end{array}$

MONDAY Patterning and Algebra

1. Write = or ≠ to make the number sentence true.

   8 − 4 ☐ 10 − 6

2. Is this a growing, shrinking, or repeating pattern?

3. Create a number pattern. ____________________

   What is your rule? ____________________

4. What is the missing number in this number pattern?

   ________ , 455, 460, 465

TUESDAY Number Sense and Operations

1. What is the value of the underlined digit?

   A. 2<u>7</u>1 ________

   B. <u>3</u>69 ________

2. Write the number.

   A. 100 less than 811 ________

   B. 100 more than 244 ________

3. How many tens and ones in 63?

   tens ________

   ones ________

4. Circle $\frac{1}{2}$ of the group.

WEDNESDAY Geometry

1. Color the cube red.
   Color the cone green.
   Color the cylinder yellow.
   Color the rectangular prism blue.

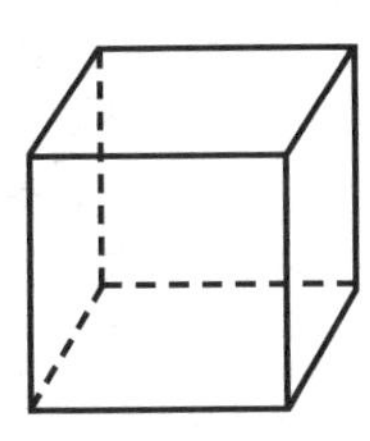
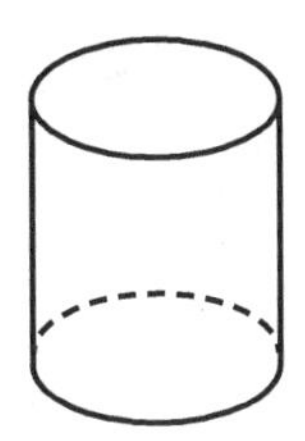
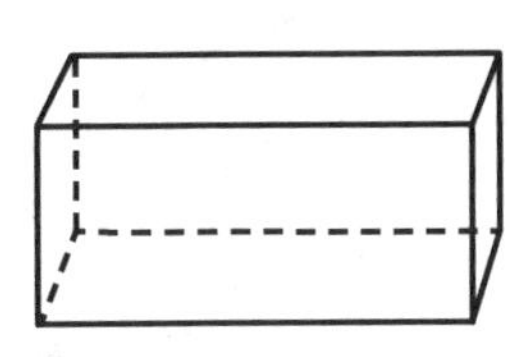
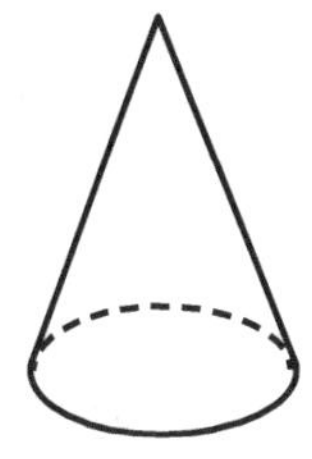

2. What is the name of this 3D shape?

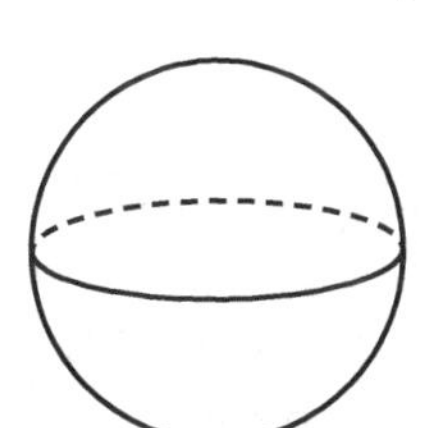

   A. sphere
   B. pyramid

3. What shape is under the circle?

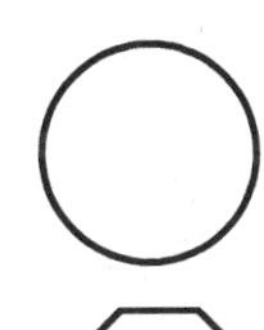

   A. octagon
   B. rectangle

THURSDAY Measurement

1. Draw in the hands on the clock to show the time 11:45.

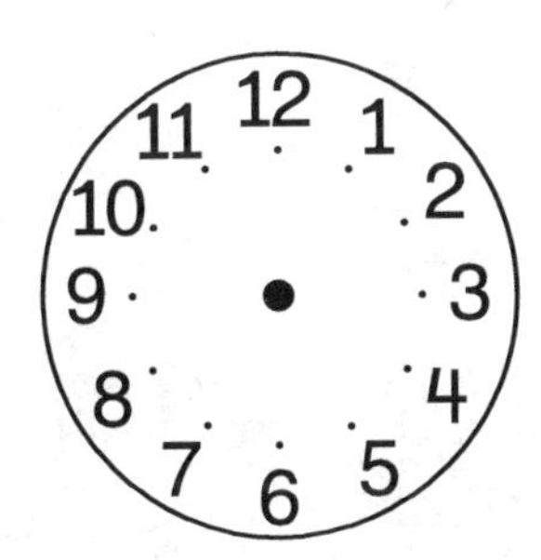

2. One cricket jumped up 9 inches. Another cricket jumped up 13 inches. How much higher did the second cricket jump?

   9 + __ = 13 ____ inches higher

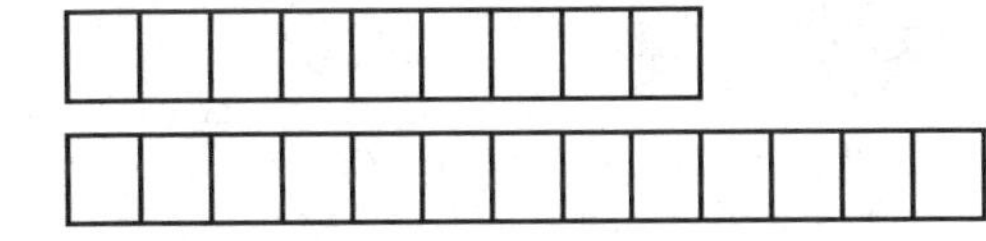

3. Choose the better unit of measure for the capacity of a large jug of juice.

   A. cup
   B. quart

4. Measure the length of the line.

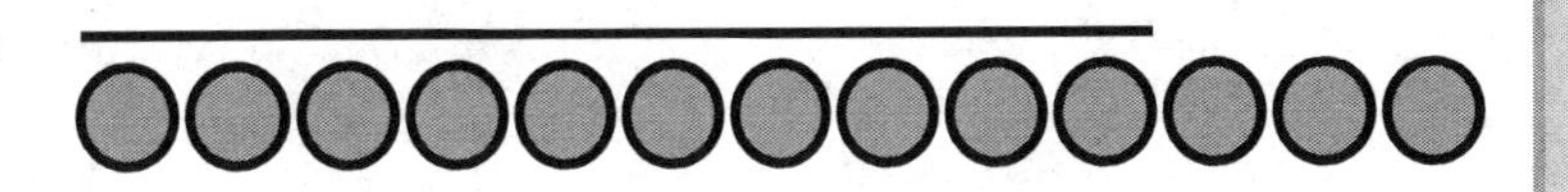

It is about __________  long.

FRIDAY Data Management

Use the Venn diagram to answer the questions about these students' favorite school clubs.

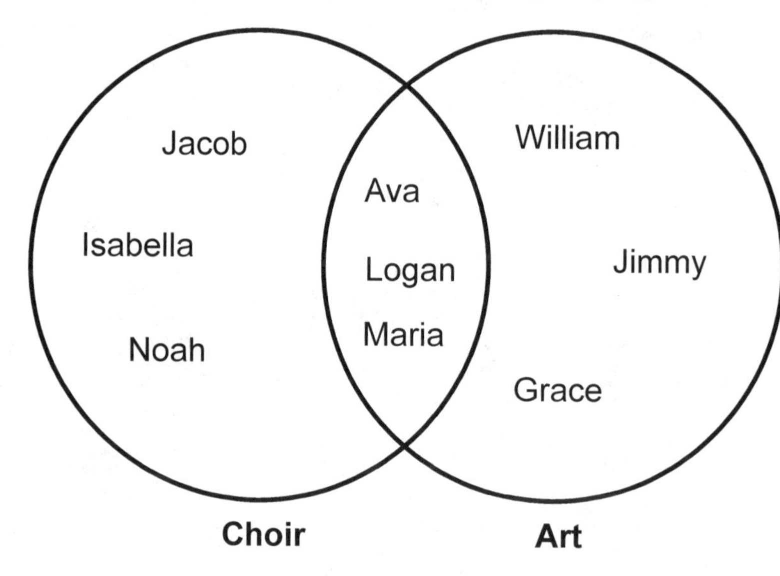

1. Which students are in the choir, but not the art club?

___

2. Which students are in the art club, but not in the choir?

___

3. How many students are in both?

___

BRAIN STRETCH

1. $641 + 264$

2. $381 + 497$

3. $564 - 327$

4. $947 - 581$

MONDAY Patterning and Algebra

1. Find the sum.

5 + 7 + 8 =

2. What is the next number if the pattern rule is subtract 10?

35, ______

3. Is this a growing, shrinking or repeating pattern?

____

4. What is the missing number in this pattern?

22, 32, 22, 32, 22, ________ , 22, 32, 22

TUESDAY Number Sense and Operations

1. What is the value of the underlined digit?

A. 57<u>8</u> ________

B. 4<u>9</u>0 ________

2. What are two related facts for 16 − 7 = 9?

_____ + _____ = _____

_____ − _____ = _____

3. Write the number in expanded form.

A. 438 = ______________________________

B. 295 = ______________________________

4. Color $\frac{1}{2}$ of the shape.

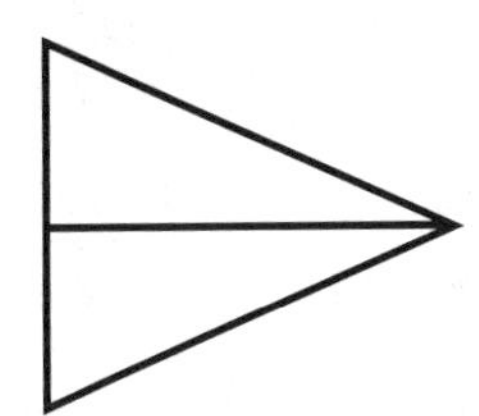

WEDNESDAY Geometry

1. Color the cylinder green.
   Color the cone red.
   Color the sphere orange.
   Color the cube blue.

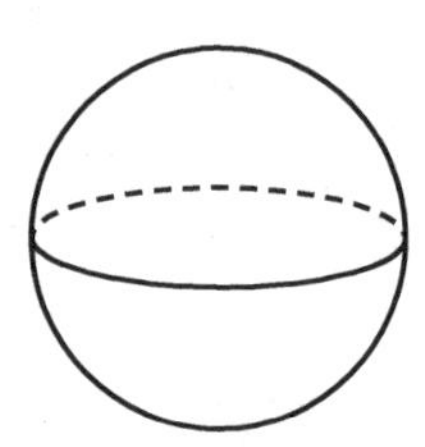
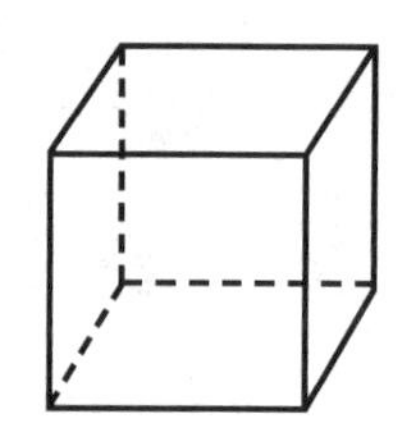
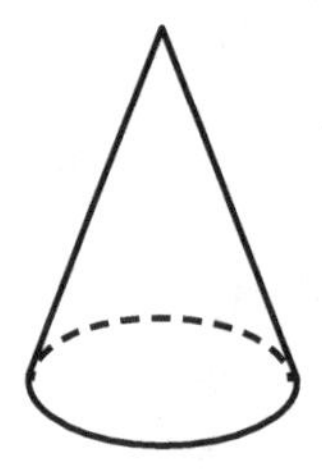
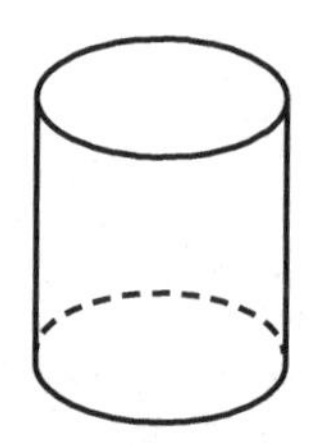

2. What is the name of this 3D shape?

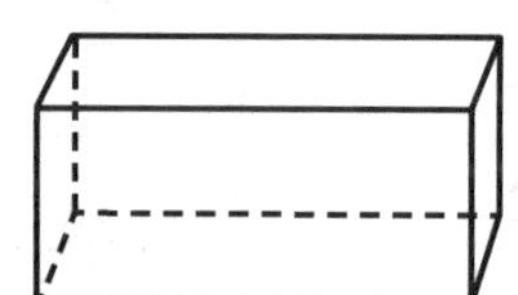

A. rectangular prism

B. cone

3. Draw a line of symmetry on the shape.

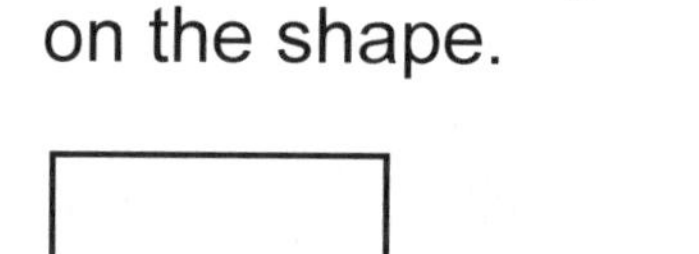

THURSDAY Measurement

1. Write the time in two ways.

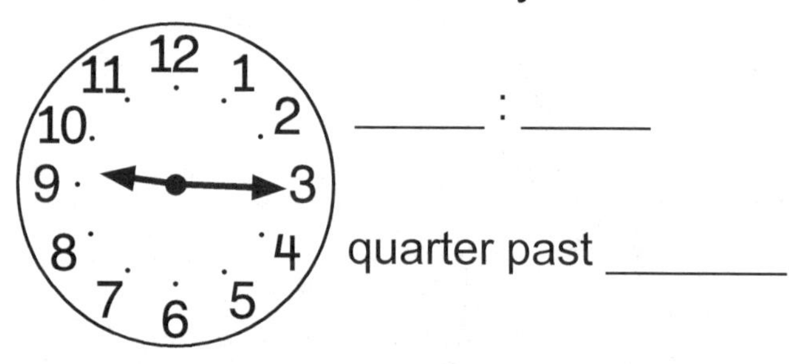

_____ : _____

quarter past _______

2. What is a better estimate of the weight of a bicycle?

A. about 1 pound

B. more than 1 pound

C. less than 1 pound

3. There were 36 puppies. 17 puppies were adopted. How many puppies are left? Use the number line.

36 – 17 = ___ ___ puppies

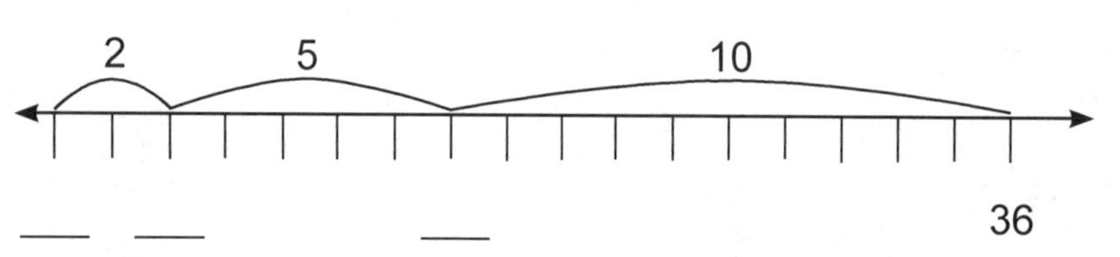

FRIDAY Data Management

Use the Venn diagram to answer the questions about these students' favorite treats.

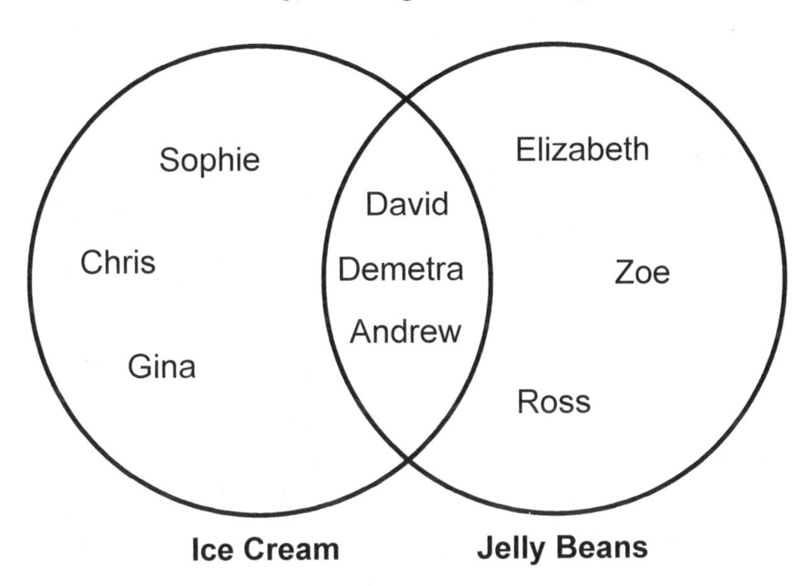

1. Which students like ice cream but not jelly beans?

____

2. How many students like jelly beans?

____

3. Which students like ice cream and jelly beans?

____

BRAIN STRETCH

1.
$$\begin{array}{r} 775 \\ +\ 184 \\ \hline \end{array}$$

2.
$$\begin{array}{r} 867 \\ +\ 129 \\ \hline \end{array}$$

3.
$$\begin{array}{r} 364 \\ -\ 192 \\ \hline \end{array}$$

4.
$$\begin{array}{r} 563 \\ -\ 214 \\ \hline \end{array}$$

MONDAY Patterning and Algebra

1. Find the difference.

$18 - 5 - 3 =$ ______

2. What is the next number if the pattern rule is add 9?

21, ______

3. Is this a growing, shrinking, or repeating pattern?

190, 180, 170, 160, 150, 140

____

4. What is the missing number in this pattern?

666, 656, 646, 636, ________ , 616, 606

TUESDAY Number Sense and Operations

1. Find the missing sign.

31 ◯ 12

A. <   B. =   C. >

2. What are two related facts for $13 - 5 = 8$?

_____ + _____ = _____

_____ − _____ = _____

3. Tell if the numbers are odd or even.

A. 19 ________________

B. 7 ________________

4. Color $\frac{1}{4}$ of the shape.

WEDNESDAY Geometry

1. Color the cylinder red. Color the cube blue. Color the pyramid green. Color the sphere orange.

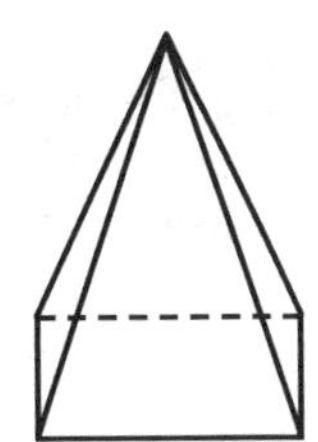 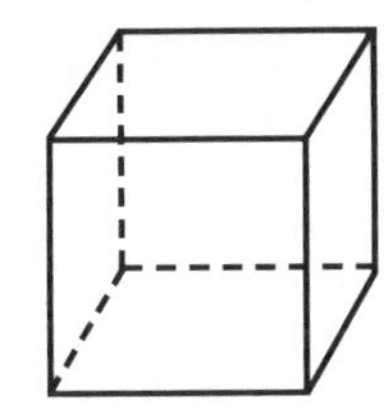 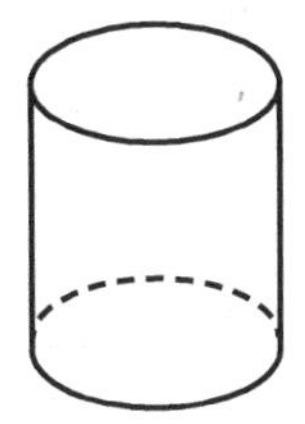 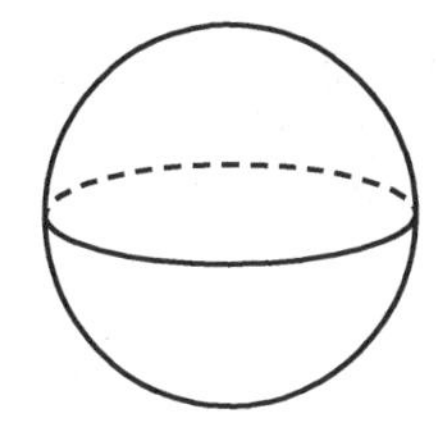

2. What is the name of this 3D shape?

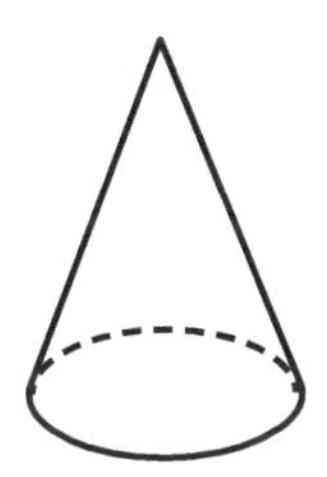

A. sphere

B. cone

3. Draw a line of symmetry on the shape.

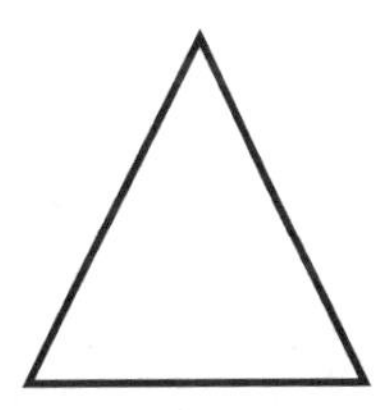

THURSDAY Measurement

1. Write the time in two ways.

_____ : _____

quarter past _______

2. How many cups in a pint?

1 pint equals _______ cups

3. What is the better estimate of the length of a bug?

A. 2 meters

B. 2 centimeters

4. Janie is 44 inches tall. Her baby sister is 18 inches tall. How much taller is Janie than her sister? Show your work.

44 – 18 = ☐

___ inches taller

FRIDAY Data Management

Use the calendar to answer the questions.

July

Sunday	Monday	Tuesday	Wednesday	Thursday	Friday	Saturday
		1	2	3	4	5
6	7	8	9	10	11	12
13	14	15	16	17	18	19
20	21	22	23	24	25	26
27	28	29	30	31		

1. How many days are there in the month of July? ____________________

2. What day of the week is July 12th? ____________________

3. What day of the week is July 21st? ____________________

4. What day of the week does the month end on? ____________________

BRAIN STRETCH

Jason needs 44¢ to buy a stamp.
If he has 17¢, how much more money does he need?

MONDAY Patterning and Algebra

1. Find the difference.

   $16 - 4 - 1 =$ ________

2. There are 9 rabbits in the garden. 11 more rabbits came. How many rabbits are in the garden? Use pictures and an equation to show your work.

   ____

   ____ rabbits

3. Color a pattern.

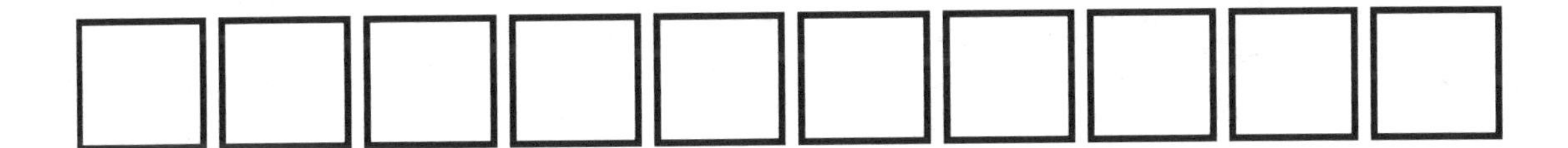

What is your pattern rule? ____

TUESDAY Number Sense and Operations

1. Find the missing sign.

   42 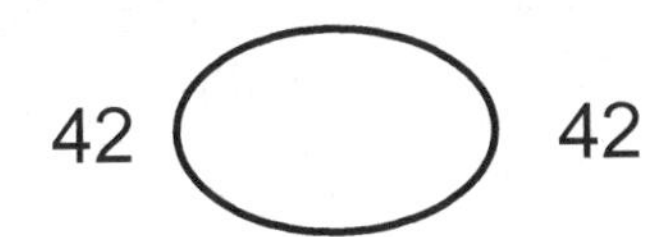 42

   A. $<$ B. $=$ C. $>$

2. Estimate the sum. Choose the better choice.

   $24 + 15 =$ ________

   A. greater than 50
   B. less than 50

3. Use the number line to help add 419 + 135. Write the equation.

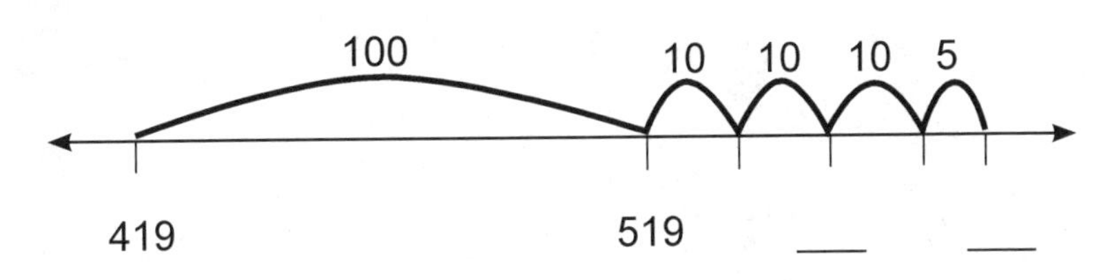

___ ___

4. Circle $\frac{1}{3}$ of the group.

WEDNESDAY Geometry

1. Color the shapes with more than 3 vertices green.

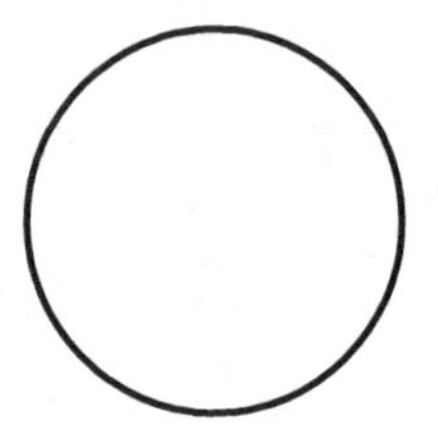 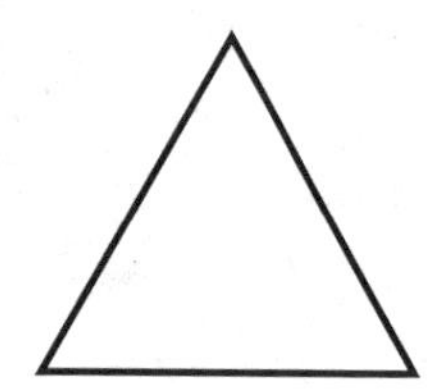 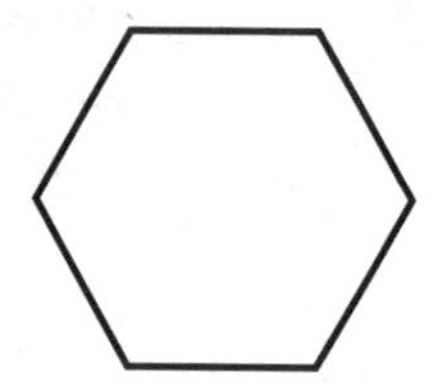  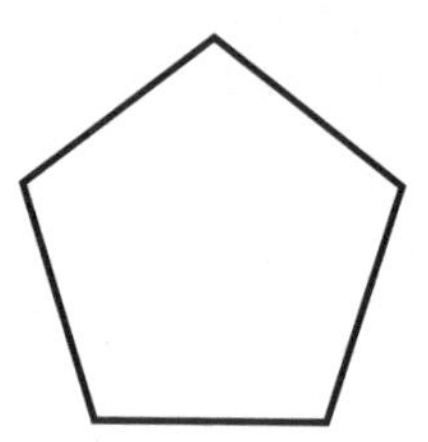

2. What is the name of this 3D shape?

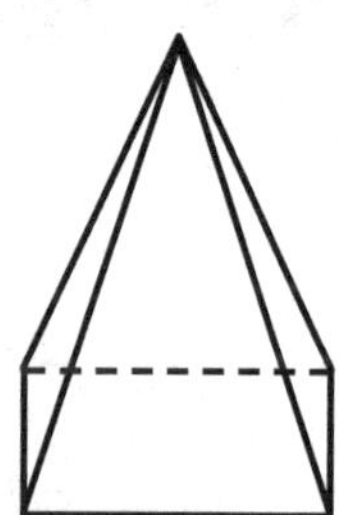

A. sphere

B. pyramid

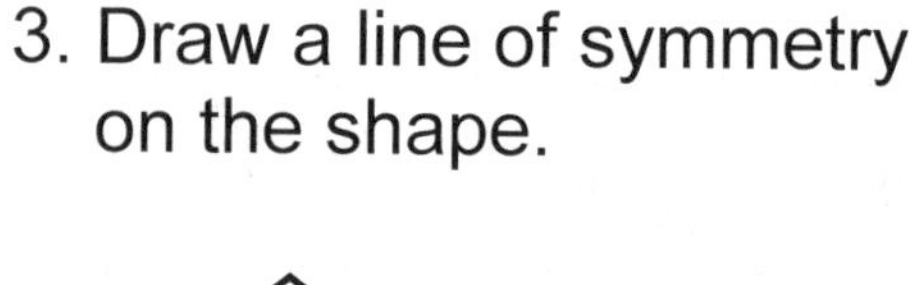

3. Draw a line of symmetry on the shape.

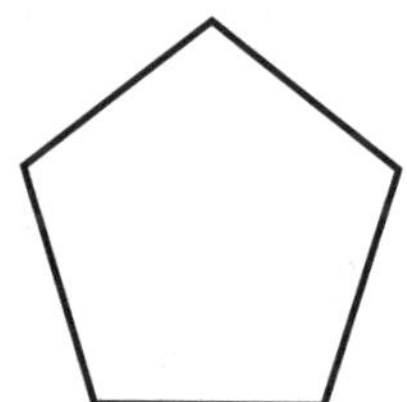

THURSDAY Measurement

1. Write the time in two ways.

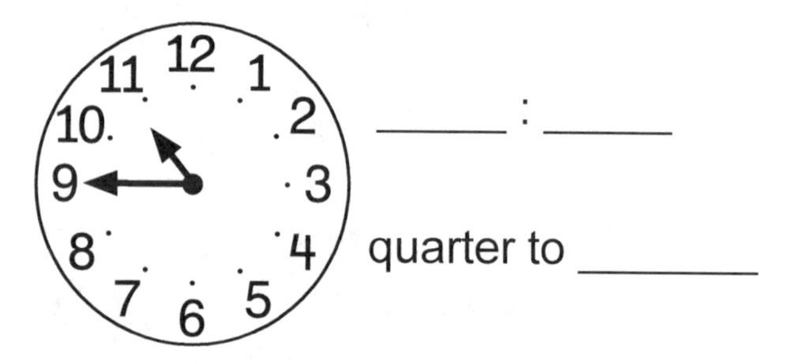

_____ : _____

quarter to _______

2. Tori went to visit her aunt from 5:00 to 8:00. How long did she visit her aunt?

_______ hours

3. What is the better estimate of the length of a car?

A. 4 meters

B. 4 centimeters

4. How many ounces in a pound?

_______ ounces

FRIDAY Data Management

Use the calendar to answer the questions.

November

Sunday	Monday	Tuesday	Wednesday	Thursday	Friday	Saturday
				1	2	3
4	5	6	7	8	9	10
11	12	13	14	15	16	17
18	19	20	21	22	23	24
25	26	27	28	29	30	

1. How many days are there in the month of November? ____________

2. What day of the week is November 16th? ____________

3. What day of the week is November 20th ____________

4. What day of the week will December start on? ____________

BRAIN STRETCH

Anna has three dogs, four cats, six hamsters and one bird.
How many pets does she have altogether?

MONDAY Patterning and Algebra

1. What is the missing number?

   ______ + 10 = 13

2. What is the next number if the pattern rule is subtract 4?

   19, ______

3. Make an AB pattern using □ and ○ .

____

TUESDAY Number Sense and Operations

1. Estimate the sum. Choose the better choice.

   28 + 51 = ______

   A. greater than 50
   B. less than 50

2. What is the sum?

$$\begin{array}{r} 693 \\ +\ 188 \\ \hline \end{array}$$

3. What is the difference?

$$\begin{array}{r} 927 \\ -\ 254 \\ \hline \end{array}$$

4. Circle $\frac{1}{4}$ of the group.

WEDNESDAY Geometry

1. Color the shapes with more than 5 vertices red.

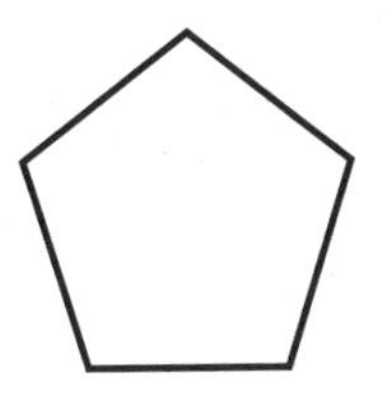 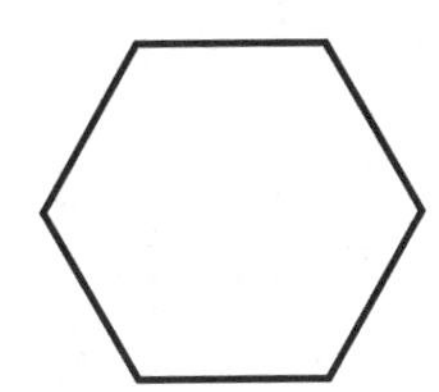 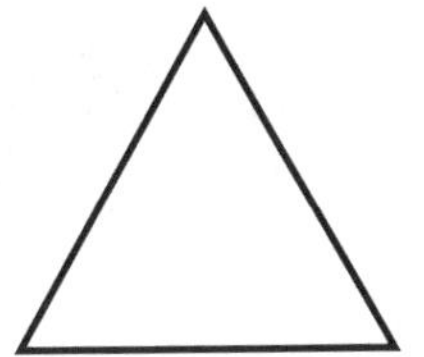 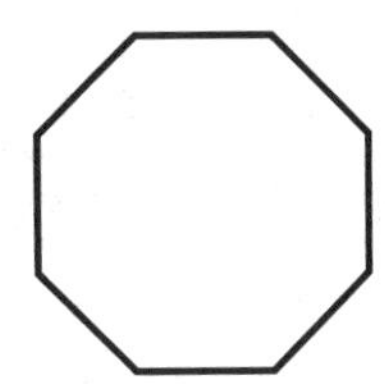 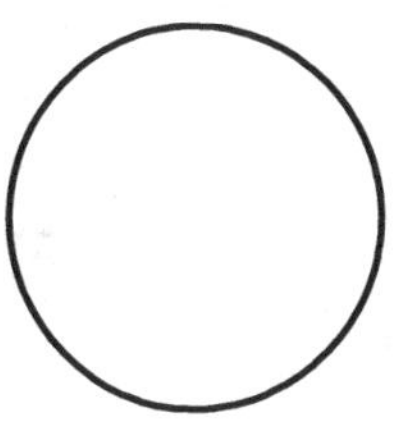

2. Can this 3D shape be stacked?

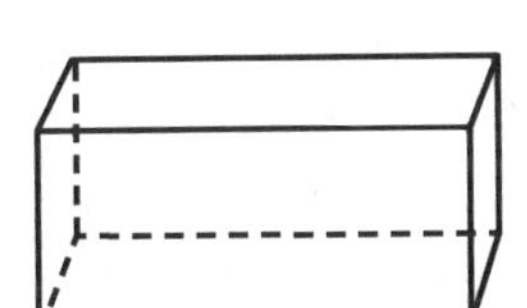

A. yes
B. no

3. Draw a line of symmetry on the shape.

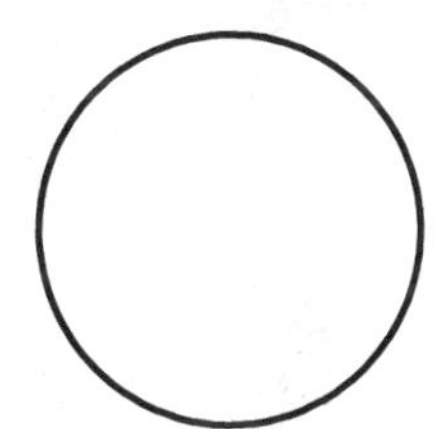

THURSDAY Measurement

1. Write the time in two ways.

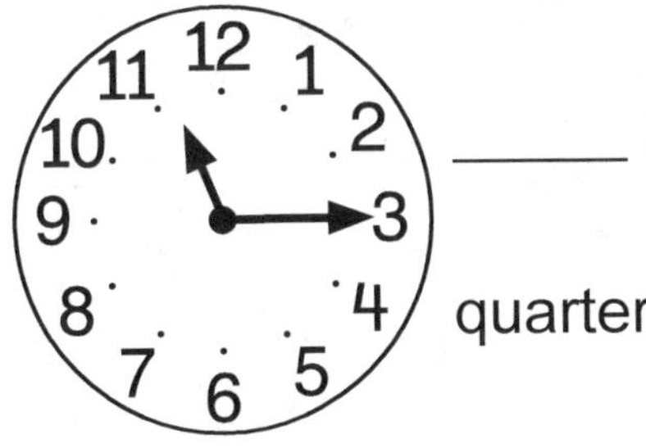

______ : ______

quarter past ________

2. How many centimeters in a meter?

There are ________ centimeters in a meter.

3. The rectangle was divided into 2 rows and 3 columns. How many squares are there?

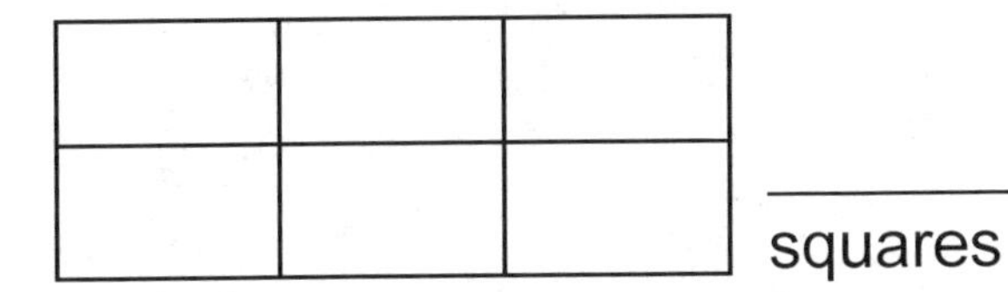

________ squares

4. Divide the rectangle. Make 2 rows and 4 columns. Check that the squares are the same size. Count the squares.

________ squares

FRIDAY Data Management

Here are the results of a Favorite Meal Survey.
Answer the questions using the information from the bar graph.

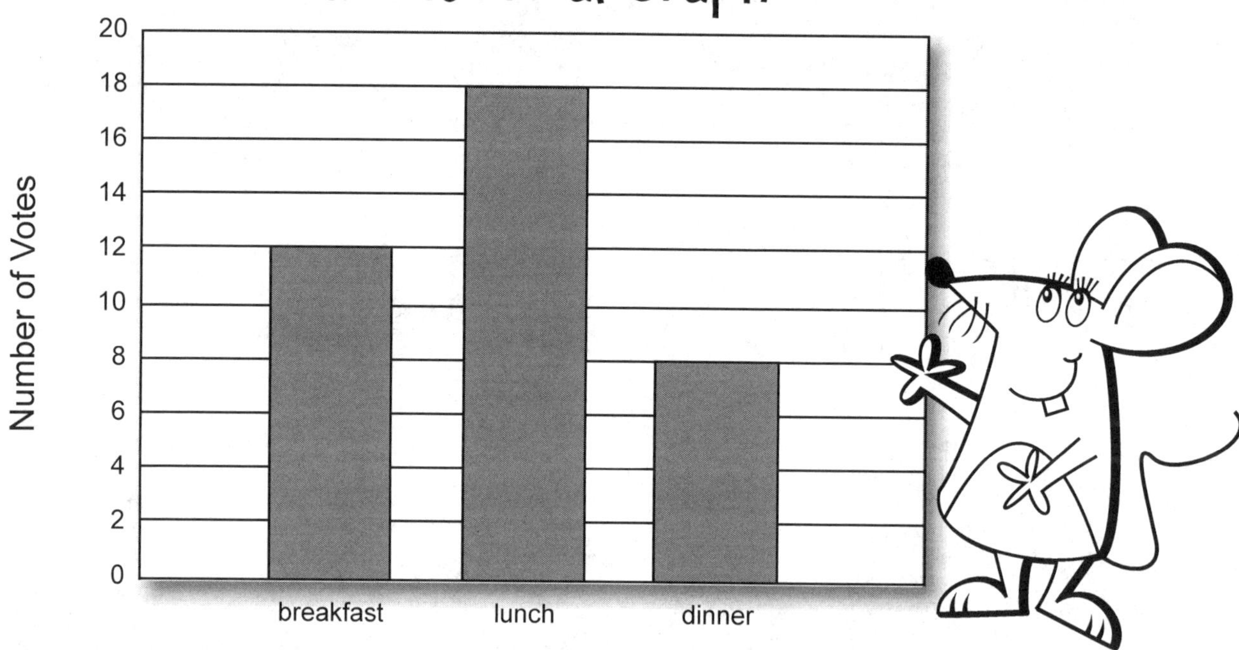

1. What is this graph about? ______________________________

2. Which meal got 8 votes? ______________________________

3. Which meal was the most popular? ______________________________

4. Which meal was the least popular? ______________________________

BRAIN STRETCH

There were 37 frogs at the pond. 19 frogs jumped away.
How many frogs were left?

MONDAY Patterning and Algebra

1. Write = or ≠ to make the number sentence true.

   1 + 11 ☐ 12 − 1

2. What is the missing number to make the equation true?

   10 + 3 = 2 + ________

3. Color a pattern.

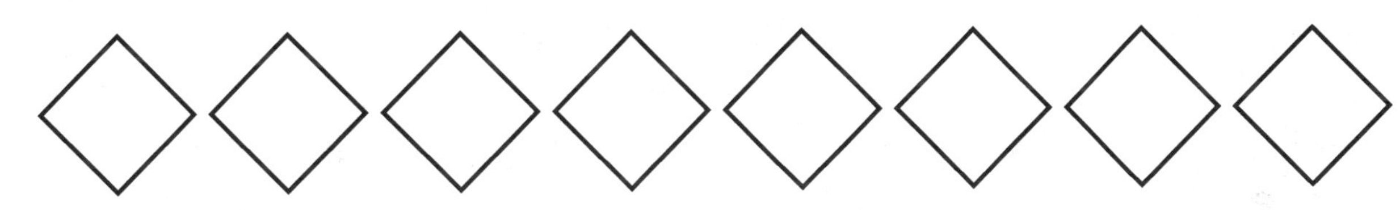

What is your pattern rule? ______________________________

4. Add or subtract.

   A. What is 10 more than 328?______

   B. What is 10 less than 328? ______

TUESDAY Number Sense and Operations

1. Compare the numbers using <, >, or =.

   366 ◯ 366

   A. <   B. =   C. >

2. Sasha solved 45 + 25.
   She added 40 + 20 + 5 + 5.

   Is her strategy correct? _____

   Show how you know.

   40 + 20 = ___   5 + 5 = ___

   60 + 10 = ___   So, 45 + 25 = ___.

3. Subtract:

   A. 600 − 100 = _________

   B. 900 − 200 = _________

4. What are two related facts for 8 + 5 = 13?

   _____ + _____ = _____

   _____ − _____ = _____

WEDNESDAY Geometry

1. Color the quadrilaterals red.

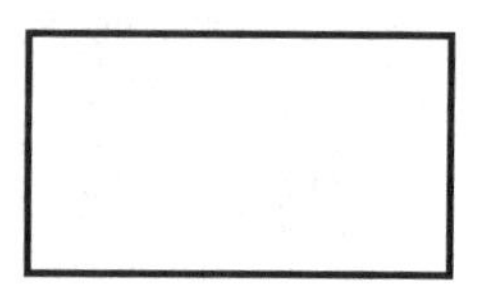
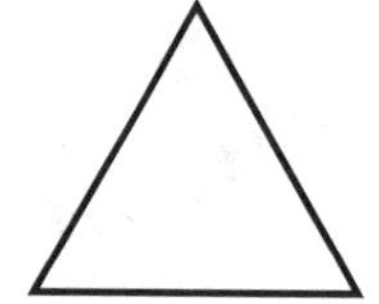
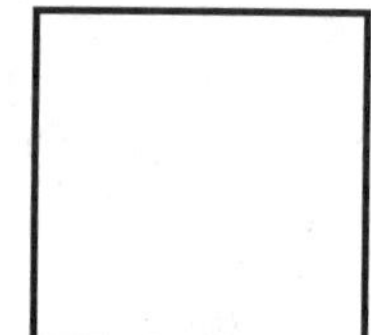
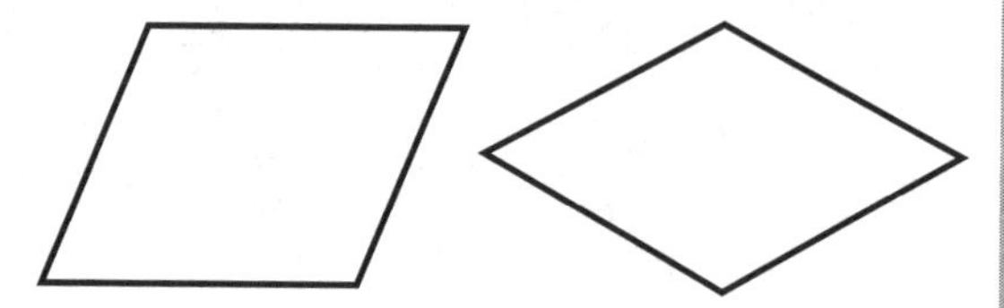

2. Can this 3D shape be stacked?

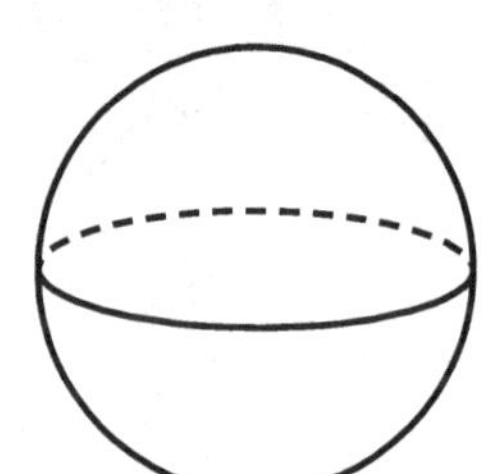

A. yes
B. no

3. Draw a line of symmetry on the shape.

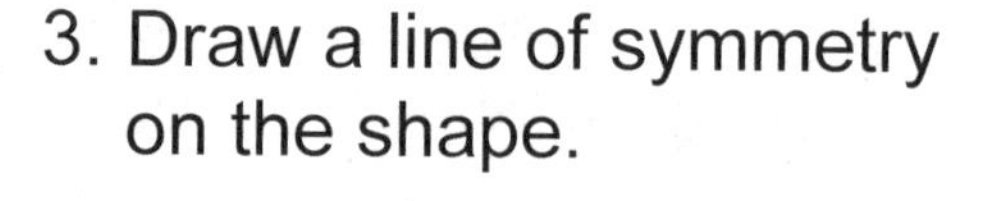

THURSDAY Measurement

1. Write the time in two ways.

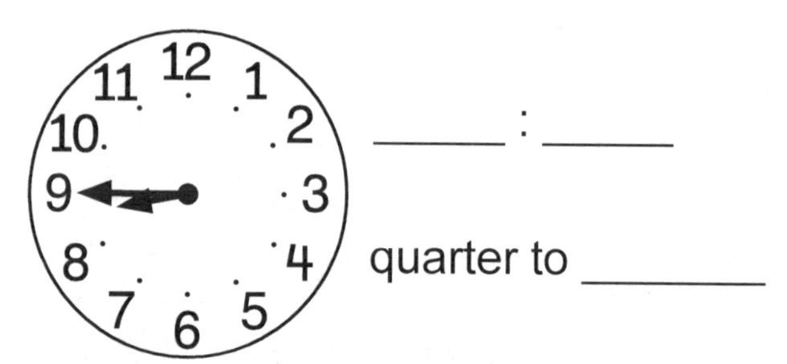

_____ : _____

quarter to _______

2. What is the better estimate of the width of a window?

A. 10 centimeters
B. 1 meter

3. Divide the rectangle.
Make 3 rows and 2 columns.
Count the squares.

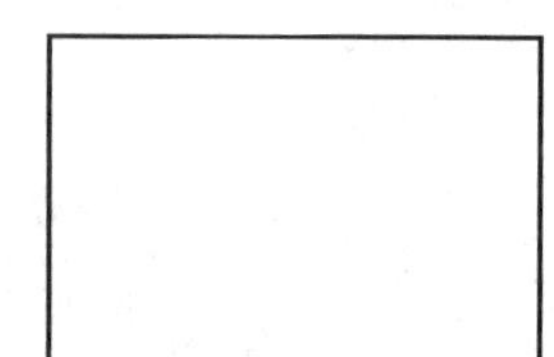

____ squares

4. Count the squares.

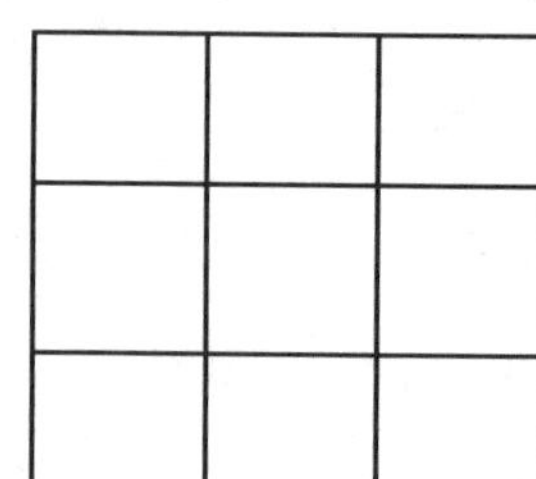

____ squares

FRIDAY Data Management

Here are the results of a Favorite Fruit Survey.
Answer the questions using the information from the bar graph.

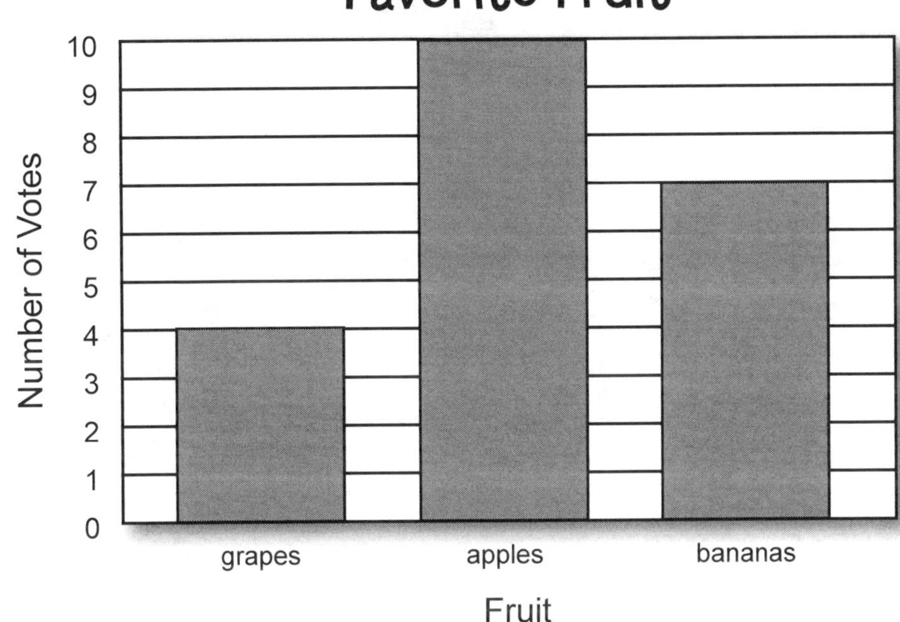

1. What fruit was the most popular? ________________
2. What fruit was the least popular? ________________
3. Order the fruits from the least number of votes to the most.

   ____
4. How many votes were there for grapes and bananas? ____________
5. How many more votes were there for apples than grapes? ____________

BRAIN STRETCH

Maria had 35 jelly beans. She bought 57 more.
How many jelly beans does Maria have altogether?

MONDAY Patterning and Algebra

1. What is the sum?

   26 + 5 = ________

2. There are 3 red cars, 12 blue cars, and 4 green cars in the parking lot. How many cars are in the lot? Draw a picture and write an equation to help you.

   ____

   ____ cars

3. What is the difference?

   67 − 8 = ________

4. Create a number pattern.

   ____

   What is your pattern rule? ______________________________

TUESDAY Number Sense and Operations

1. Compare the numbers using <, >, or =.

   433 ◯ 433

   A. < B. = C. >

2. Freddie is solving 25 + 16. He uses place value blocks. He has 3 tens and 11 ones. 11 ones is like 1 ten and 1 one. So that makes 4 tens and 1 one. What does that make? ___

   Write the equation.

   ____

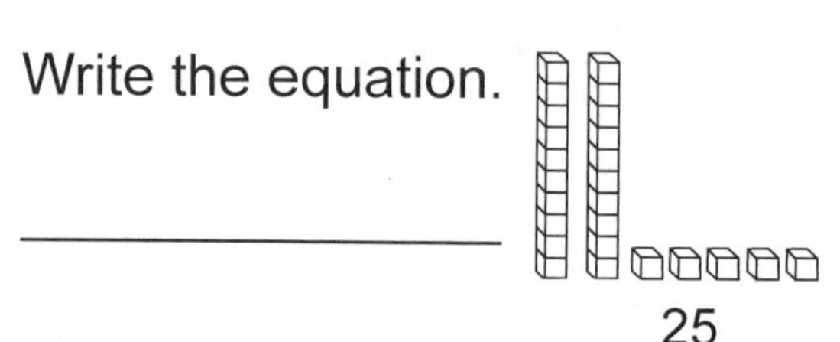

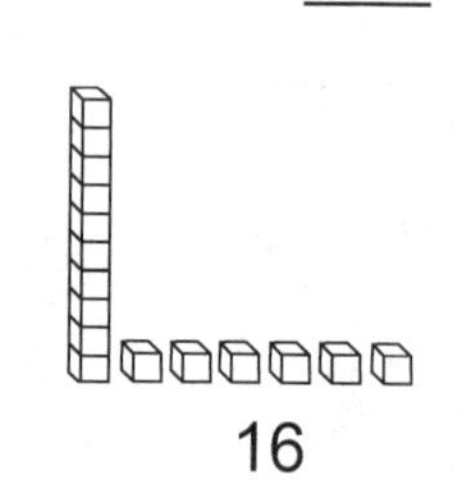

3. Write the numbers.

   A. seven hundred thirty ________

   B. two hundred four ________

   C. five hundred eleven ________

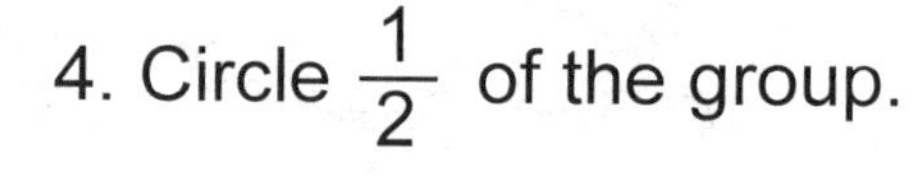

4. Circle $\frac{1}{2}$ of the group.

WEDNESDAY Geometry

1. Color the shapes with fewer than 6 vertices green.

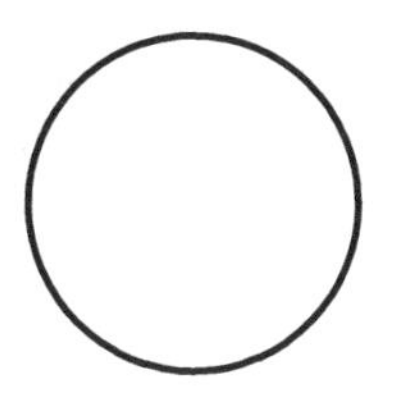 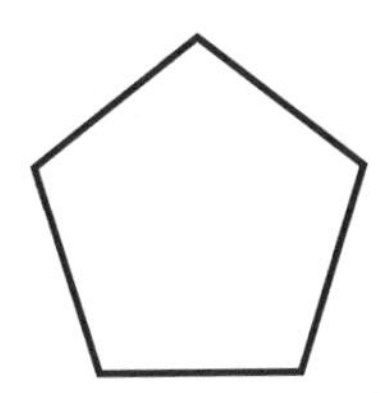 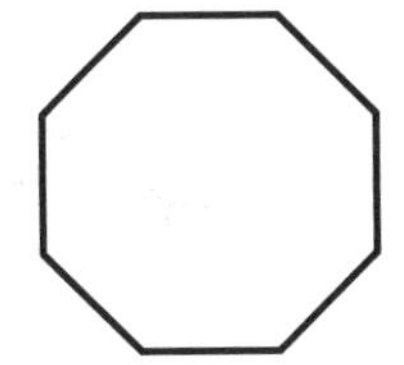  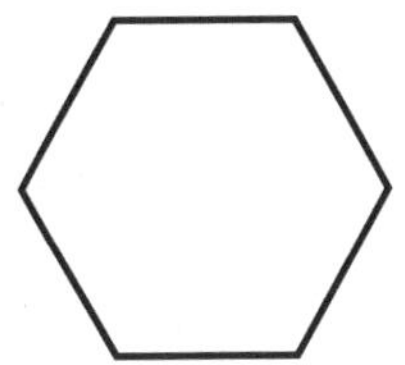

2. Can this 3D shape roll?

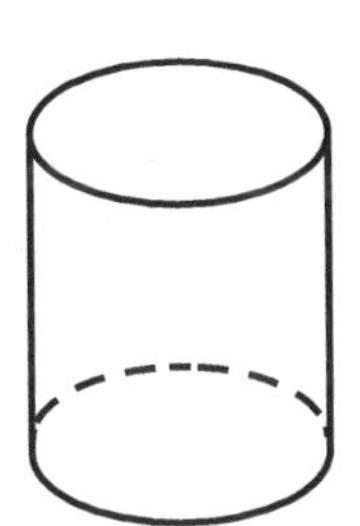

A. yes
B. no

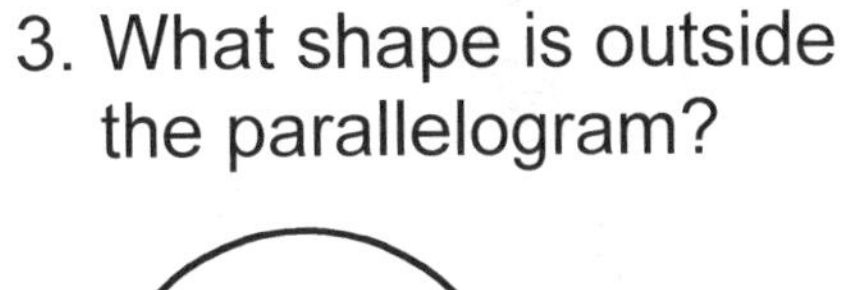

3. What shape is outside the parallelogram?

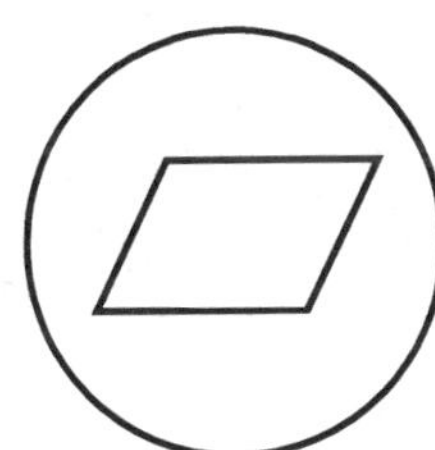

A. pentagon
B. circle

THURSDAY Measurement

1. What time will it be in 2 hours?

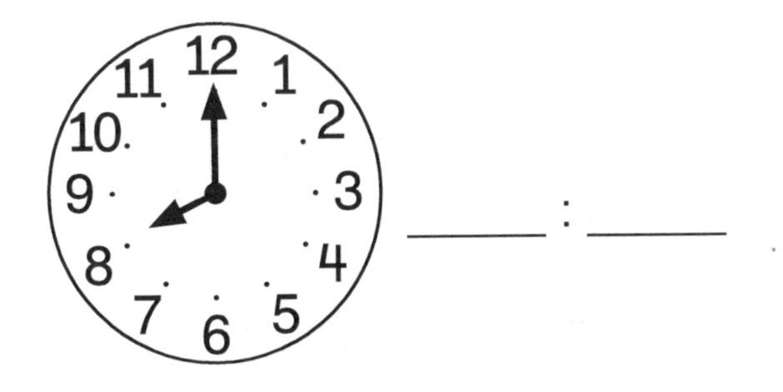

_____ : _____

2. Which holds more?

A. 2 pints
B. 2 cups

3. The rectangle has __ equal parts.
There are __ fourths altogether.
Each part is called one _______.

_______ units

4. How many squares are there?

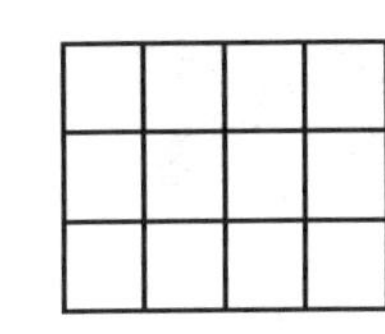

_______ squares

FRIDAY Data Management

Here are the results of a Favorite Music Survey.
Answer the questions using the information from the bar graph.

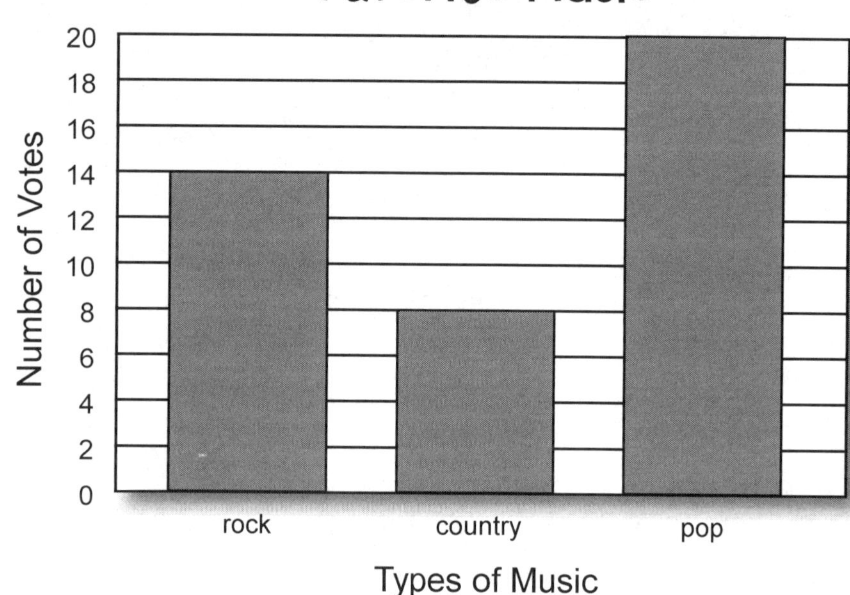

1. What music was the most popular? ________________

2. What music was the least popular? ________________

3. Order the types of music from the most number of votes to the least.

____

4. How many votes were there for rock and country? ________________

5. How many fewer votes were there for country than pop? ____________

BRAIN STRETCH

Chris had 67 hockey cards. He gave 19 cards to Stephen.
How many hockey cards does Chris have left?

MONDAY Patterning and Algebra

1. What is the sum?

33 + 9 = ________

2. What is the difference?

81 − 5 = ________

3. Count on by 10s.

540, 550, ______, ______, ______, ______, ______

4. What is the missing number in this number pattern?

114, _________, 134, 144, 154

TUESDAY Number Sense and Operations

1. Use the number line to choose the number that completes the sentence.

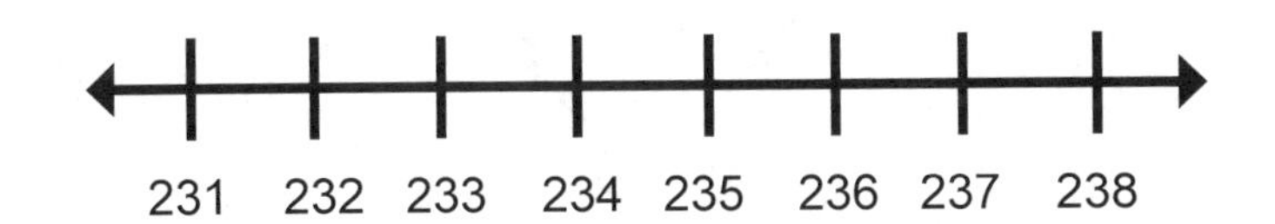

236 is two more than _________

2. List the numbers from greatest to least.

285, 176, 390

______ > ______ > ______

3. Add.

A. 200 + 300 = ________

B. 60 + 30 = ________

4. Circle the fourth unicorn.

WEDNESDAY Geometry

1. Color the octagons red.
   Color the pentagons yellow.
   Color the trapezoid green.
   Color the rectangles blue.

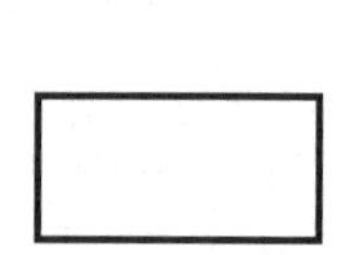 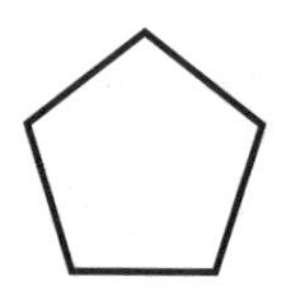  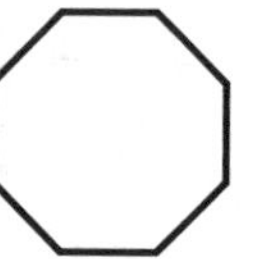 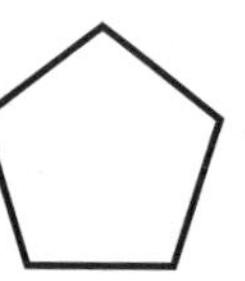  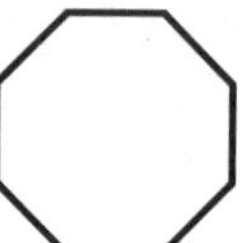

2. Can this 3D shape roll?

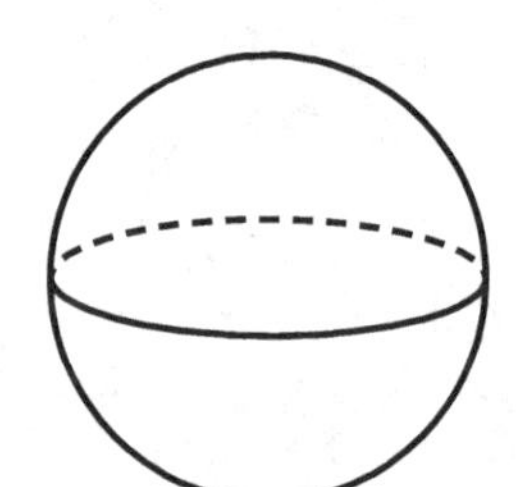

A. yes
B. no

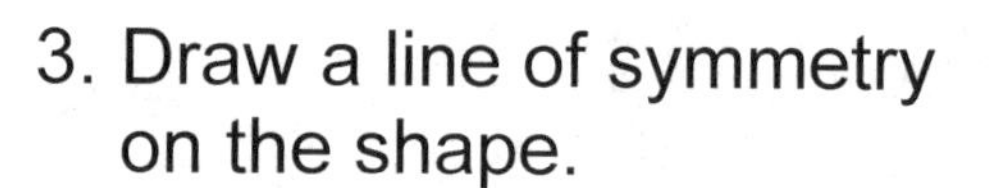

3. Draw a line of symmetry on the shape.

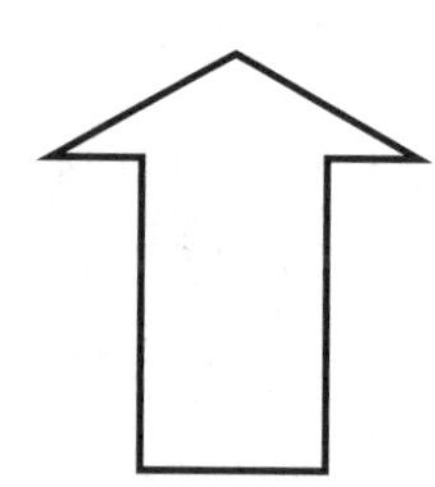

THURSDAY Measurement

1. Draw in the hands on the clock to show the time 10:25.

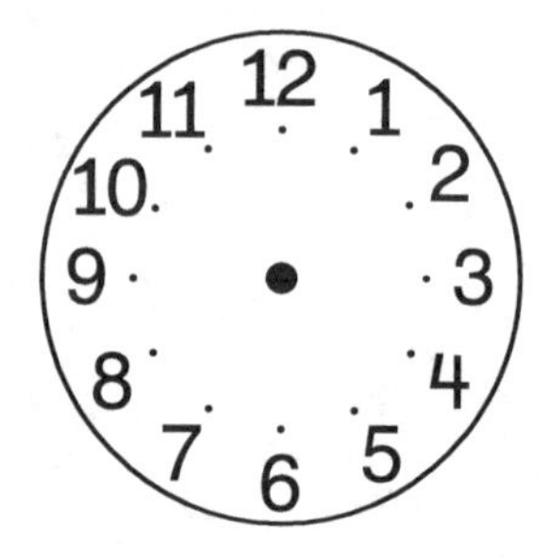

2. What month is just after September?

A. December
B. October
C. April

3. Count the squares.

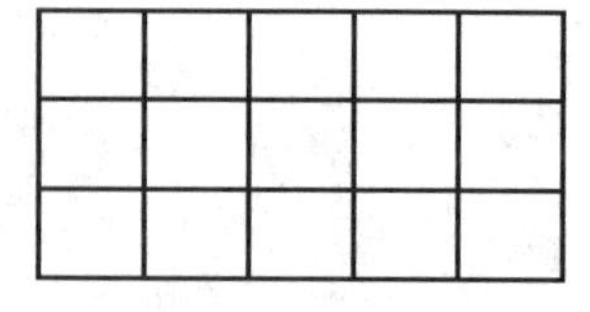

_______ squares

4. Divide the rectangle into 2 equal parts.
   There are ___ halves altogether.
   Each part is called one ______.

FRIDAY Data Management

Here are the results of a Favorite School Subject Survey.
Use the data from the chart to make a bar graph. Answer the questions.

Favorite School Subject

Subject										
Art										
Language										
Math										
Science										
Music										
	0	1	2	3	4	5	6	7	8	9	10

Number

Favorite School Subject

School Subject	Number
Art	6
Language	9
Math	8
Science	8
Music	4

1. What was the most popular subject? ____________________

2. How many voted for art and math? ______________

3. Which subjects had same number of votes? __________________________

BRAIN STRETCH

Cathy baked 24 chocolate chip cookies and 36 oatmeal raisin cookies.
How many cookies did Cathy bake in all?

MONDAY Patterning and Algebra

1. Count on by 100s from 222.

   222, ________, ________, ________, ________

2. What is the difference?

   35 − 9 = ________

3. Jason has 9 oranges. Lucy has 17 oranges. How many more oranges does Lucy have than Jason. Use pictures and an equation to show your work.

   ____

   Lucy has ___ more oranges than Jason.

4. What is the missing number in this number pattern?

   517, 527, ________ , 547, 557

TUESDAY Number Sense and Operations

1. Use the number line to choose the number that completes the sentence.

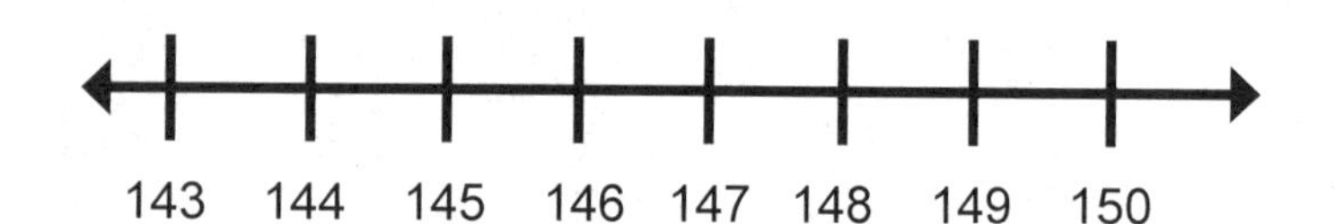

   148 is one less than ________

2. Find the sum.
   Write the equation.

   6 groups of 2

   ○○ ○○ ○○

   ○○ ○○ ○○

   ____

3. Subtract.

   A. 800 − 700 = ________

   B. 600 − 400 = ________

4. Circle the sixth elephant.

WEDNESDAY Geometry

1. Look at the shapes.
   Choose flip, slide, or turn.

A. flip   B. slide   C. turn

2. Look at the shapes.
   Choose flip, slide, or turn.

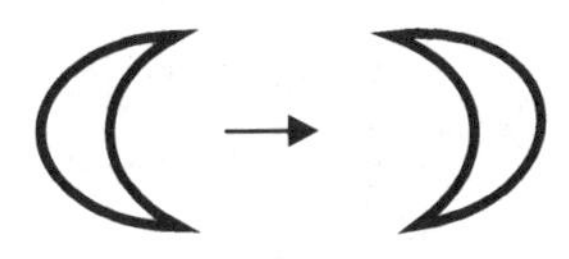

A. flip   B. slide   C. turn

3. Can this 3D shape roll?

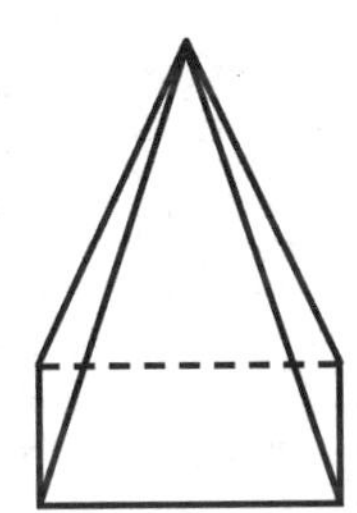

A. yes
B. no

4. Draw a line of symmetry
   on this letter.

THURSDAY Measurement

1. Write the time in two ways.

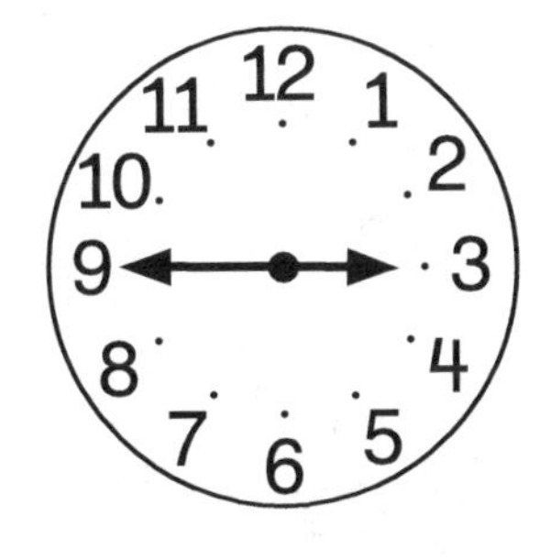

_____ : _____

quarter to _______

2. Which is lighter?

A. 

B. 

3. What length is longer?

A. 6 inches
B. 6 feet

4. What is the area?

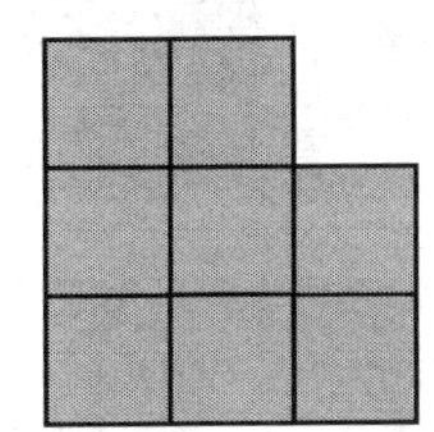

_______ square units

FRIDAY Data Management

Here are the results of a Favorite Fair Ride Survey.
Answer the questions using the information from the bar graph.

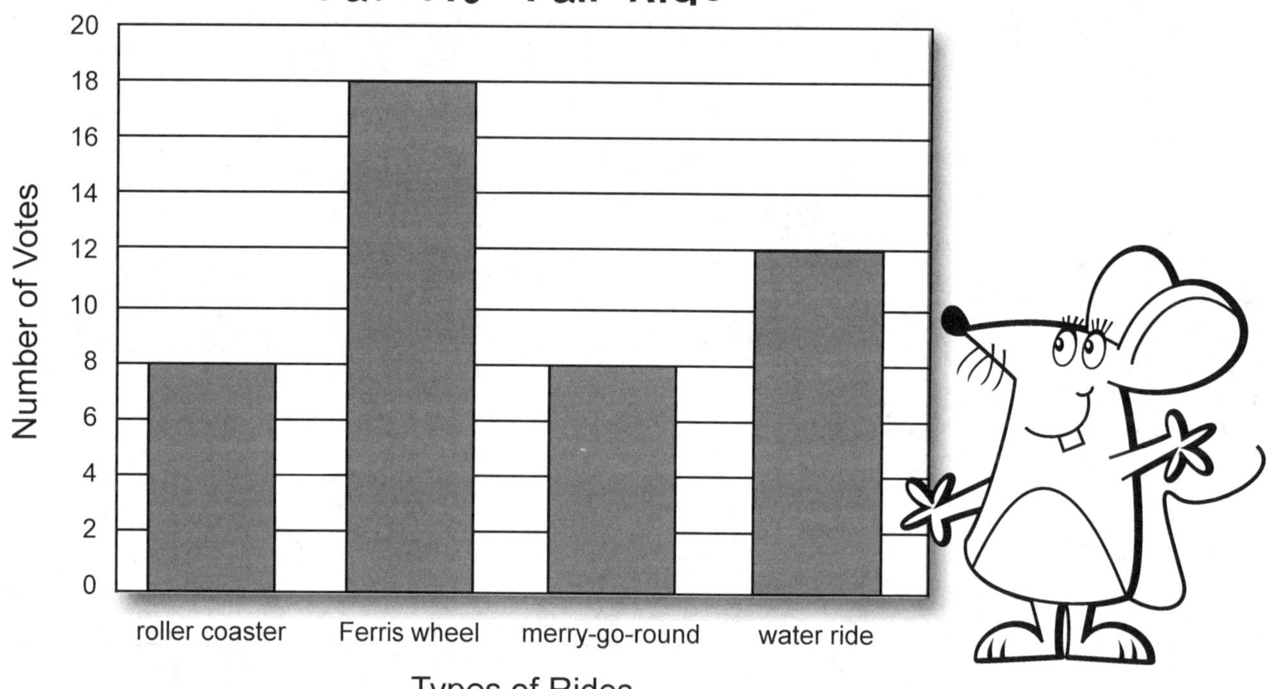

1. Which ride was the most popular? ___________________________

2. Which two rides had the same number of votes ? ________________________

3. How many fewer votes for the water ride than the roller coaster? _________

4. How many votes for the merry-go-round? ___________

BRAIN STRETCH

There were 81 ants on a log. 57 of the ants were eaten by an anteater. How many ants were left on the log?

MONDAY Patterning and Algebra

1. Count on by 5s from 885.

885, _____, _____, _____, _____

2. What is the difference?

61 − 9 = _______

3. Count on by 5s from 565.

565, 570, _____, _____, _____, _____, _____

4. Write = or ≠ to make the number sentence true.

10 + 10 ☐ 18 − 1

TUESDAY Number Sense and Operations

1. Find the sum.
   Write the equation.

○ ○ ○ ○ ○
○ ○ ○ ○ ○

____

2. Estimate the sum.
   Choose the better choice.

67 + 32 = _______

A. greater than 100
B. less than 100

3. Subtract.

A. 900 − 500 = _______
B. 300 − 100 = _______

4. What are two related facts for 14 − 5 = 9?

_____ + _____ = _____

_____ − _____ = _____

WEDNESDAY Geometry

1. Color the hexagons red. Color the circles blue.
   Color the triangles yellow. Color the parallelogram green.

2. Can this 3D shape roll?

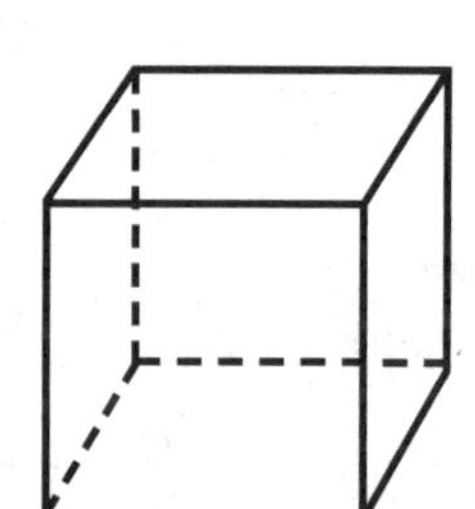

A. yes
B. no

3. What shape is above the parallelogram?

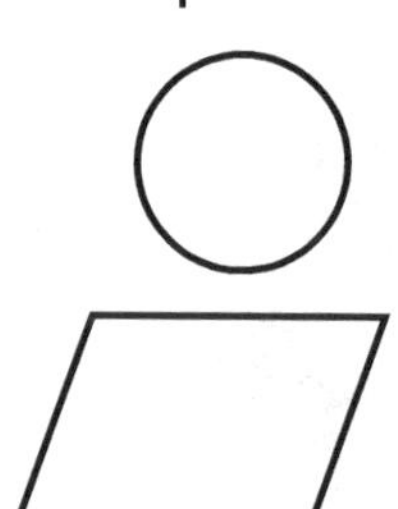

A. pentagon
B. circle

THURSDAY Measurement

1. Draw in the hands on the clock to show the time 2:40.

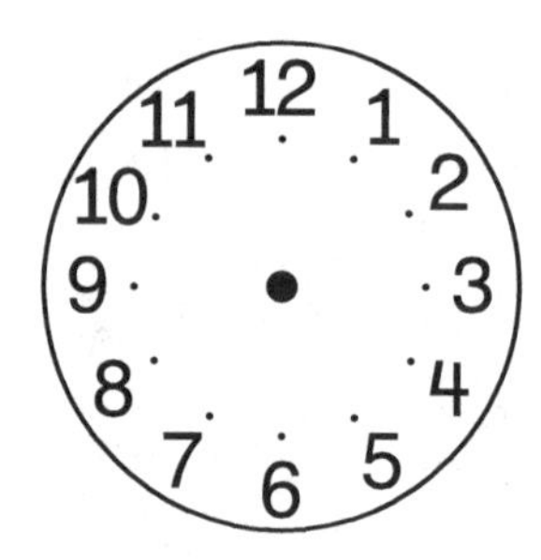

2. Divide the rectangle into 3 equal parts. There are __ thirds altogether. Each part is called one ______.

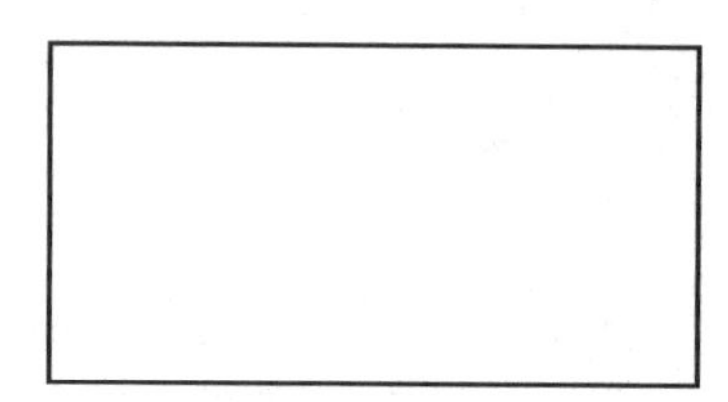

3. Draw a line 5 inches long.

4. Divide the circle into 2 equal parts. There are ___ halves altogether. Each part is called one ____.

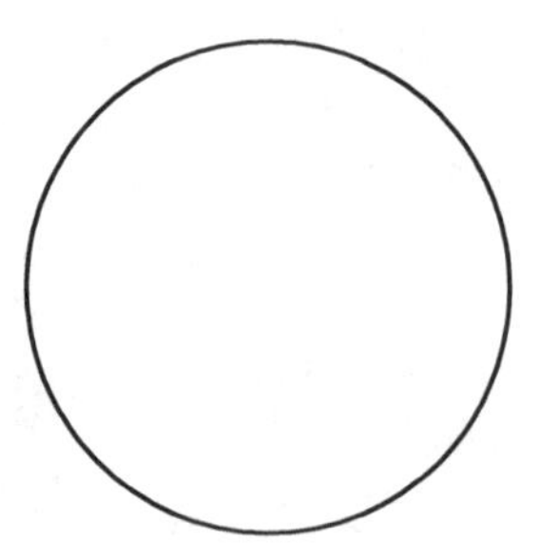

FRIDAY Data Management

Here are the results of a Favorite Color Survey.
Answer the questions using the information from the tally chart.

Favorite Color

Color	Tally	Number
Red	~~||||~~ ||||	
Blue	~~||||~~ ~~||||~~ |||	
Green	~~||||~~ |||	
Yellow	||	

1. What was the most popular color? ______________________

2. What was the least popular color? ______________________

3. How many people liked either green or yellow? ______________

4. Order the colors from the most number of votes to the least.

____

BRAIN STRETCH

Carlos has 5 dimes. He trades them with Mia for nickels.
How many nickels does Carlos now have?

MONDAY Patterning and Algebra

1. What is the sum?

45 + 6 = ________

2. What is the difference?

72 − 4 = ________

3. Mentally add 100 to each number.

A. 135 _____ B. 348 _____ C. 541 _____

4. What is the missing number to complete the number sentence?

________ − 4 = 7 + 3

TUESDAY Number Sense and Operations

1. Find the sum. Write the addition sentence.

○ ○ ○ ○ ○
○ ○ ○ ○ ○
○ ○ ○ ○ ○
○ ○ ○ ○ ○

____

2. Estimate the sum. Choose the better choice.

49 + 56 = ________

A. greater than 100
B. less than 100

3. What are two related facts for 3 + 9 = 12?

______ + ______ = ______

______ − ______ = ______

4. Color $\frac{3}{4}$ of the shape.

WEDNESDAY Geometry

1. Color the pentagons red.
   Color the triangle blue.
   Color the octagons orange.
   Color the trapezoids green.

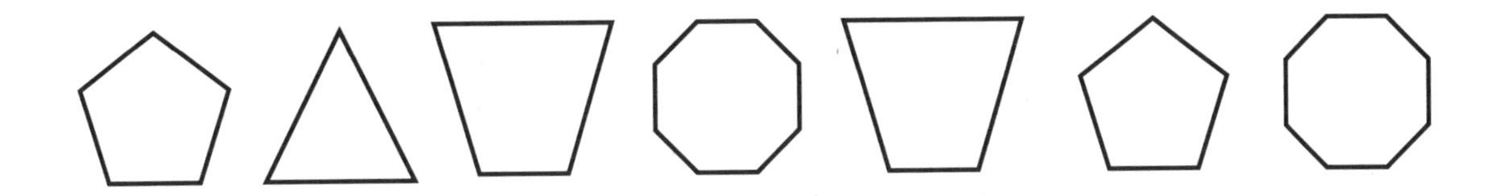

2. Look at the shapes.
   Choose flip, slide, or turn.

A. flip B. slide C. turn

3. Draw a line of symmetry on this letter.

THURSDAY Measurement

1. What time was it 1 hour ago?

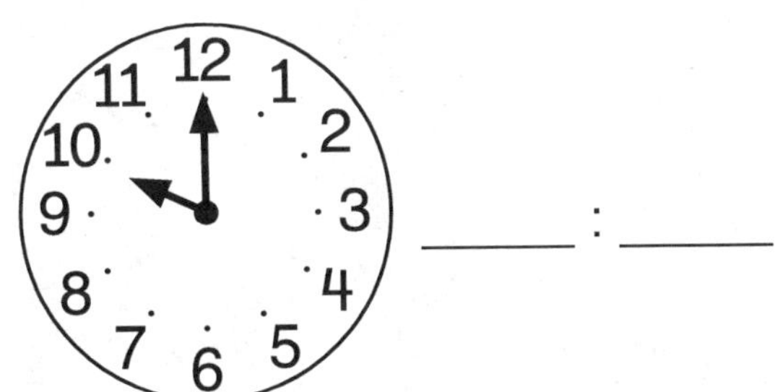

_____ : _____

2. What could be the temperature on a hot day?

____

3. Susie has 3 quarters and 2 dimes. How many cents does she have? Use pictures and words.

___ cents

4. Compare the lengths.
   Which length is longer than the line?

A.

B.

FRIDAY Data Management

Use the Venn diagram to answer the questions about these students' favorite circus performers.

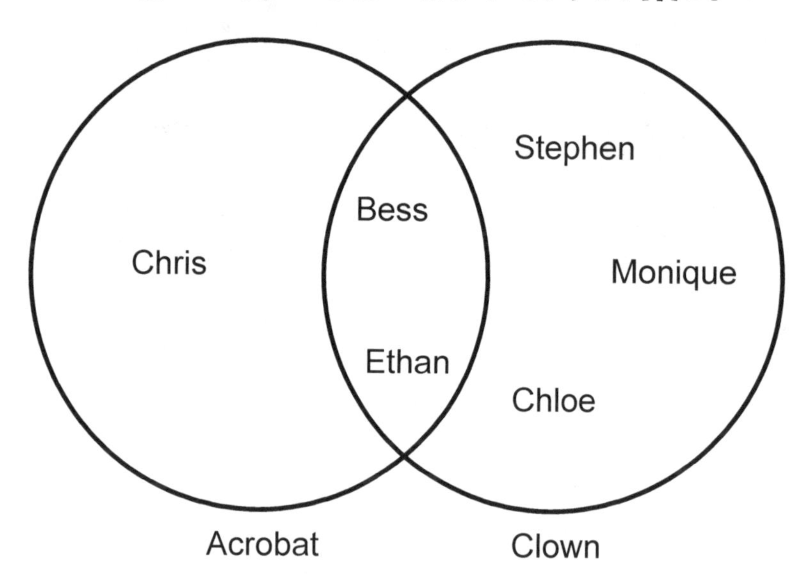

1. Which students like clowns, but not acrobats?

____

2. Which students like acrobats?

____

3. Which students like clowns and acrobats?

____

BRAIN STRETCH

Lisa bought 3 boxes of cupcakes. Each box has 4 cupcakes.
How many cupcakes does Lisa have altogether?

MONDAY Patterning and Algebra

1. What is the sum?

22 + 8 = ________

2. What is the difference?

43 − 10 = ________

3. Create a number pattern.

____

What is your pattern rule?

____

4. There were some students in the gym. Then 14 more students came. There are now 37 students. How many students were in the gym at the start?

Use pictures and an equation to show your work.

____

___ students

TUESDAY Number Sense and Operations

1. Find the sum. Write the addition sentence.

____

○ ○ ○ ○ ○
○ ○ ○ ○ ○
○ ○ ○ ○ ○

2. Write the numeral.

A. four hundred and eighty six ________

B. nine hundred and twelve ________

C. one hundred and two ________

3. Add.

A. 300 + 300 = ________

B. 200 + 500 = ________

4. Add. Show your work.

25 + 50 + 43 + 32 = ____

WEDNESDAY Geometry

1. Color the shapes that are the same size and shape.

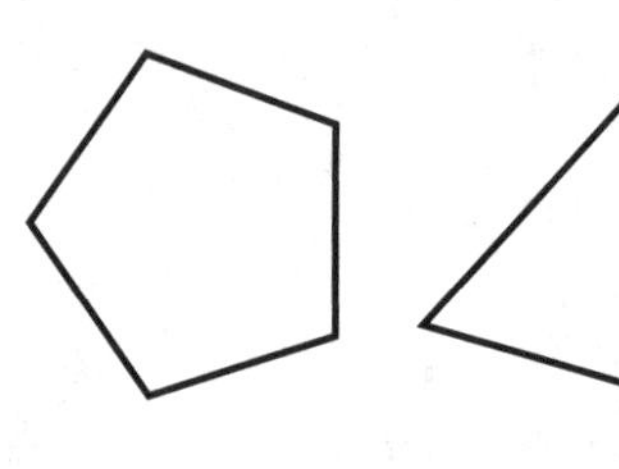
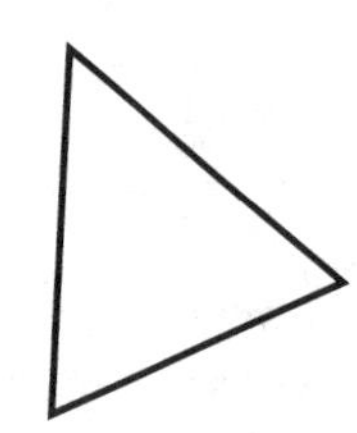
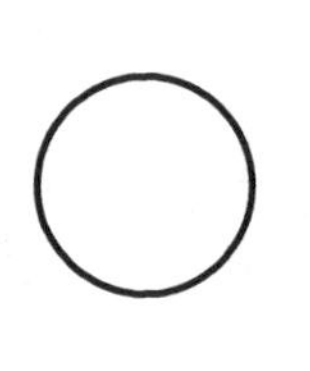
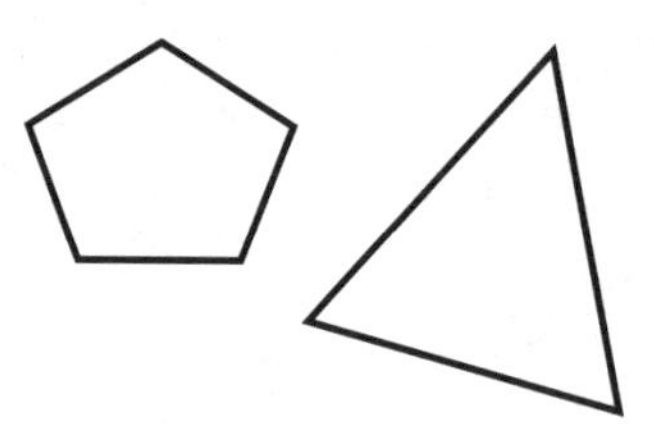

2. Look at the shapes.
   Choose flip, slide, or turn.

A. flip B. slide C. turn

3. How many sides does this shape have?

THURSDAY Measurement

1. What time will it be 1 hour later?

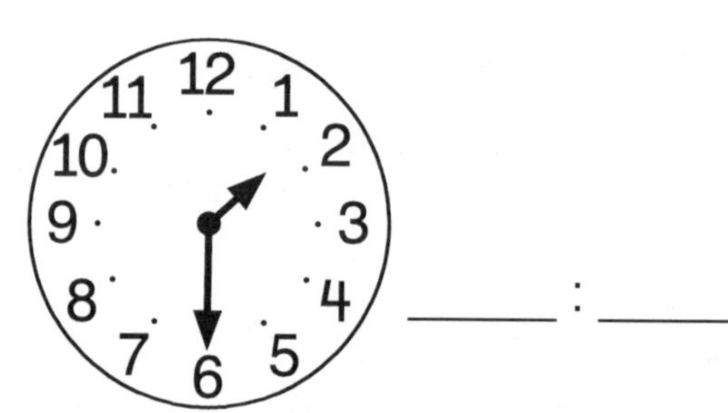

____ : ____

2. Which set holds more?

3 pints

or

2 quarts

3. If you have 4 dimes and 3 nickels, how many cents do you have? Use pictures and words.

___ cents

4. Compare the lengths.
   Which length is longer than the line?

A.

B.

FRIDAY Data Management

Here are the results of a favorite transportation survey.
Complete the tally chart and answer the questions.

Favorite Transportation

	Number	Tally
	9	
	17	
	13	

1. Circle the most popular transportation.   

2. Circle the least popular transportation.   

3. How many people liked  more than  ? ____________

BRAIN STRETCH

Bill wants to buy a fishing rod for $1.00. He has 92¢.
How much more money does Bill need to buy the fishing rod?

MONDAY Patterning and Algebra

1. What is the missing number to make the equation true?

   6 + 4 + 7 = 2 + _______

2. What is the next number if the pattern rule is subtract 2?

   22, ________

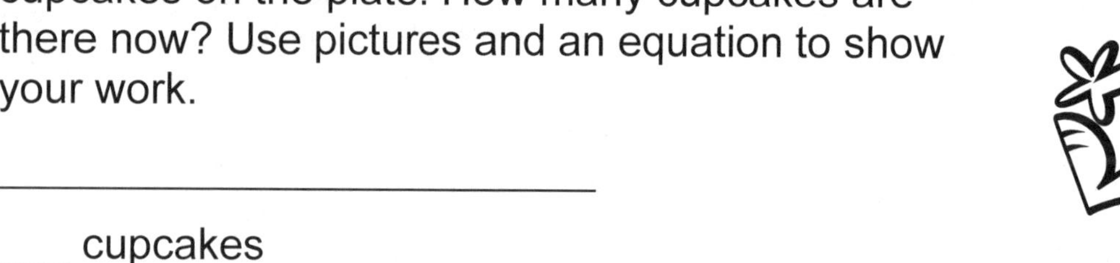

3. There are 9 cupcakes on a plate. Maria and her friends ate 4 cupcakes. Then her mom put 7 more cupcakes on the plate. How many cupcakes are there now? Use pictures and an equation to show your work.

   ____

   ___ cupcakes

TUESDAY Number Sense and Operations

1. Find the sum. Write the equation.

   ○○
   ○○
   ○○

   ____

2. Write the numeral.

   A. ninety seven _______

   B. five hundred sixty _______

   C. seven hundred eleven _______

3. Add.

   A. 200 + 40 + 6 = _______

   B. 2 + 4 + 6 = _______

   C. 20 + 46 = _______

4. Add. Show your work.

   34 + 45 + 53 + 21 = ___

WEDNESDAY Geometry

1. Color the shapes that are the same size and shape.

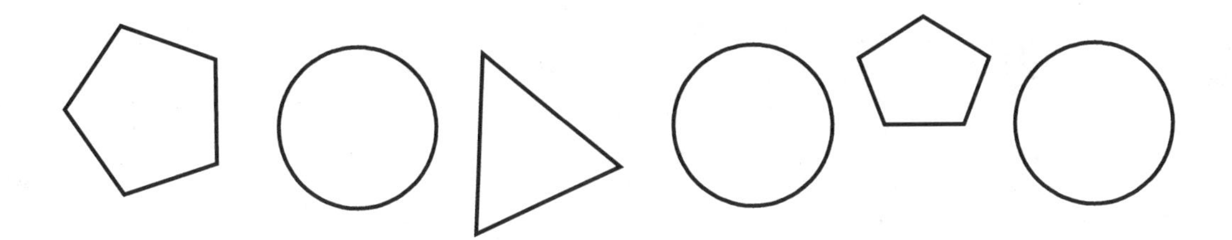

2. Look at the shapes.
   Choose flip, slide, or turn.

A. flip   B. slide   C. turn

3. How many vertices does this shape have?

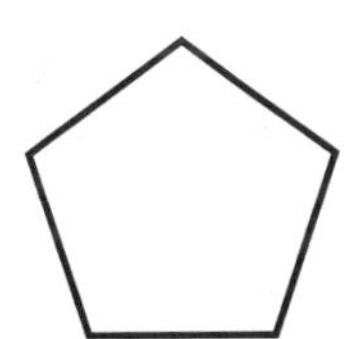

THURSDAY Measurement

1. Write the time in two ways.

____ : ____

quarter to ______

2. Draw a line 3 inches long.

3. If you have 1 quarter and 2 dimes, how many cents do you have?

___ cents

4. Divide the circle into 3 equal parts. There are ___ thirds altogether.Each part is called ____ -____.

FRIDAY Data Management

Ben went fishing. Look at the chart to see the number of fish Ben caught from Monday to Friday.

Number of Fish Ben Caught

Day of the Week	Monday	Tuesday	Wednesday	Thursday	Friday
Number of Fish Caught	2	4	6	8	10

1. On what day did Ben catch the most number of fish? ____________________

2. On what day did Ben catch the least number of fish? ____________________

3. What is the difference between the most number of fish Ben caught and the least number of fish? ____________________

BRAIN STRETCH

Complete the chart. Show three different ways to make $1.00 with coins.

Ways to Make $1.00 Using Coins

	Quarter	Dime	Nickel	Penny
$1.00				
$1.00				
$1.00				

Week 1, pages 1–3

Monday **1.** Accept any repeating pattern with a matching explanation. **2.** 4 **3.** 22, 24, 26

Tuesday **1.** 6 tens, 5 ones, number 65 **2.** 16 **3.** 10 **4.** A

Wednesday **1.** triangle **2.** 3 **3.** 3

Thursday **1.** 7:00 **2.** cold **3.** A **4.** 3

Friday **1.** 5 **2.** 4 **3.** **4.**

Brain Stretch 9 balloons

Week 2, pages 4–6

Monday **1.** Accept any repeating pattern. **2.** 14 **3.** 80, 90, 100

Tuesday **1.** 4 tens, 8 ones, number 48 **2. A.** 40 **B.** 13 **3.** 15 **4.** C

Wednesday **1.** circle **2.** 0 **3.** 0

Thursday **1.** B **2.** B **3.** A **4.** 1

Friday **1.** 8 **2.** 5 **3.** 3 **4.** 16

Brain Stretch 6 ants

Week 3, pages 7–9

Monday **1.** Accept any repeating pattern with a matching rule. **2.** Sample answer: 6 + 4 = 10; 10 + 8 = 18
3. 600, 700, 800

Tuesday **1.** 5 hundreds, 5 tens, 1 ones, number 551 **2. A.** 5 hundreds, 7 tens, 4 ones
B. 6 hundreds, 8 tens, 2 ones **3.** **4.** B

Wednesday **1.** square **2.** 4 **3.** 4

Thursday **1.** 6:00 **2.** A **3.** B **4.** 4

Friday **1.** 16 **2.** circle **3.** square **4.** triangle, rectangle

Brain Stretch 9 pieces of bubble gum

Week 4, pages 10–12

Monday **1.** The colored numbers line up vertically and skip a column, in an AB pattern. **2.** 70

Tuesday **1.** 6 hundreds, 4 tens, 0 ones, number 640 **2. A.** > **B.** = **3.** **4.** B

Wednesday **1.** rectangle **2.** 4 **3.** 4

Thursday **1.** 1:30 **2.** B **3.** A **4.** 7

Friday **1.** 8 **2.** 3 **3.** 7 **4.** 18

Brain Stretch 32 stamps

Week 5, pages 13–15

Monday **1.** The colored numbers line up vertically and end in 5 or 0. **2.** 85

Tuesday **1.** 4 tens, 3 ones, number 43 **2.** **3.** $200 < 346 < 647$ **4. A.** 66 **B.** 62

Wednesday **1.** pentagon **2.** 5 **3.** 5

Thursday **1.** 8:30 **2.** 12 inches **3.** C **4.** 2

Friday **1.** 8 **2.** 16 **3.** 10 **4.** 6

Brain Stretch 7 apples

Week 6, pages 16–18

Monday	**1.** The numbers line up vertically in a single row and all end in 0. **2.** 70
Tuesday	**1.** 1 hundred, 7 tens, 3 ones, number 173 **2. A.** 901 **B.** 111 **3. A.** 5 hundreds, 7 tens, 4 ones **B.** 6 hundreds, 8 tens, 2 ones **4.** B
Wednesday	**1.** hexagon **2.** 6 **3.** 6
Thursday	**1.** A **2.** A **3.** B **4.** 5
Friday	The Favorite Zoo Animal Graph should extend shading to: 6 for Lion, 2 for Zebra, 5 for Flamingo, and 3 for Giraffe **1.** 16 **2.** lion **3.** zebra **4.** 8
Brain Stretch	7 stamps

Week 7, pages 19–21

Monday	**1. 1.** 12, 15, 18 **2.** 13 **3.** 13 **4.** striped, solid, empty; ABC
Tuesday	**1.** 3 hundreds, 8 tens, 5 ones, number 385 **2.** ≠ **3. A.** 552 **B.** 538 **4.** 26¢
Wednesday	**1.** octagon **2.** 8 **3.** 8
Thursday	**1.** The long hand should point to 12 and the short hand to 5. **2.** 2 hours **3.** Lines should measure 1 inch. **4.** A
Friday	**1.** 7 **2.** 4 **3.** 6 **4.** 17 **5.** oranges
Brain Stretch	**1.** 15 **2.** 13 **3.** 9 **4.** 9

Week 8, pages 22–24

Monday	**1.** 105, 110, 115, 120 **2.** 17 **3.** 7 **4.** ● ▲ ●; AB
Tuesday	**1.** 5 hundreds, 6 tens, 4 ones, number 564 **2.** ≠ **3. A.** 80 **B.** 35 **4.** 20¢
Wednesday	**1.** parallelogram **2.** 4 **3.** 4
Thursday	**1.** 12:30 **2.** 4 hours **3.** A **4.** Lines should measure 3 cm.
Friday	The Favorite Pets Chart tally column: Dog 卌 IIII; Cat 卌 II; Hamster IIII; Bird II The Favorite Pets Graph should extend shading to 9 for dog, 7 for cat, 4 for hamster, and 2 for bird. **1.** dog **2.** bird **3.** 11 **4.** 3
Brain Stretch	**1.** 13 **2.** 20 **3.** 6 **4.** 9

Week 9, pages 25–27

Monday	**1.** 65 **2.** 8 + 5 = 13. So 13 – 8 = 5. **3.** □ ■; AB
Tuesday	**1.** 2 hundreds, 4 tens, 5 ones, number 245 **2.** < **3. A.** 335 **B.** 490 **4.** 36¢
Wednesday	**1.** trapezoid **2.** 4 **3.** 4
Thursday	**1.** 2:25 **2.** A **3.** Friday **4.** B
Friday	The Favorite Cookie Chart Tally Column: Chocolate Chip 卌 卌 II; Oatmeal Raisin 卌 III; Gingerbread IIII **1.** 12 **2.** 8 **3.** 8 **4.** chocolate chip
Brain Stretch	**1.** 25 **2.** 30 **3.** 48 **4.** 23

Week 10, pages 28–30

Monday	**1.** 52, add ten **2.** 6 **3.** 900 + 70 + 8, 800 + 40 + 2, 700 + 30 + 1, 600 + 20 + 5 **4.** 8
Tuesday	**1.** 5 hundreds, 5 tens, 3 ones, number 553 **2.** 2 + 8 = 10, 10 – 2 = 8 or 10 – 8 = 2 **3. A.** 800 + 40 + 2 **B.** 200 + 90 + 5 **4.** 46¢
Wednesday	**1.** cylinder **2.** 2 **3.** 3 **4.** C
Thursday	**1.** 4:40 **2.** 10:30 **3.** 12 months **4.** B
Friday	The Favorite Shape Graph should extend shading to 4 for Hexagon, 3 for Heart, and 7 for Circle. **1.** circle **2.** heart **3.** 14 **4.** 4
Brain Stretch	**1.** 17 **2.** 14 **3.** 27 **4.** 31

Week 11, pages 31–33

Monday **1.** 6 **2.** 9 **3.** ; ABB

Tuesday **1.** 9 hundreds, 3 tens, 2 ones, number 932 **2.** No; even 

3. Three of the four parts should be colored. **4.** 91¢

Wednesday **1.** cube **2.** 12 **3.** 6 **4.** ○ **5.** A

Thursday **1.** half past 2 **2.** A **3.** B **4.** A

Friday **1.** broccoli **2.** 5 **3.** carrots **4.** 11

Brain Stretch **1.** 94 **2.** 60 **3.** 30 **4.** 32

Week 12, pages 34–36

Monday **1.** 3 **2.** 7 – 4 + 8 = 11; 11 **3.** 930, 940, 950, 960 **4.** 0

Tuesday **1.** 3 hundreds, 9 tens, 6 ones, number 396 **2.** 341 < 351 **3. A.** 451 **B.** 380 **4.** 80¢

Wednesday **1.** cone **2.** 1 **3.** 2 **4.** Sample answer: △ **5.** A

Thursday **1.** 8:15, quarter past 8 **2.** A **3.** C **4.** 2

Friday **1.** 30 **2.** Monday **3.** 5 **4.** Friday

Brain Stretch **1.** 61 **2.** 110 **3.** 25 **4.** 57

Week 13, pages 37–39

Monday **1.** 1 **2.** 3 + 3 + 3 + 3 = 12 **3.** Accept any pattern with a matching rule.

Tuesday **1.** 2 hundreds, 6 tens, 6 ones, number 266 **2.** 2; 287 < 289

	Hundreds	Tens	Ones
287	2	8	7
289	2	8	9

3. A. 263 **B.** 569 **4.** 20 cents

Wednesday **1.** rectangular prism **2.** 12 **3.** 6 **4.** □ **5.** A

Thursday **1.** 4:15, quarter past 4 **2.** B **3.** A **4.** A

Friday Bar graphs should be shaded to 5 for spring, 10 for summer, 2 for autumn, and 9 for winter.

1. 15 **2.** 4 **3.** autumn **4.** 19

Brain Stretch **1.** 62 **2.** 83 **3.** 59 **4.** 31

Week 14, pages 40–42

Monday **1.** + **2.** 16 **3.** circles, heart, circle, heart **4.** 90, 92, 94, 96

Tuesday **1.** □□□ IIIIII ••• **2.** odd; There is one circle left over so 5 is odd. 

3. equal; equal; 4; 562 < 566

	Hundreds	Tens	Ones
562	5	6	2
566	5	6	6

4. 82¢

Wednesday **1.** sphere **2.** 0 **3.** 0 **4.** ▭ **5.** A

Thursday **1.** 12:15, quarter past 12 **2.** 3 feet **3.** 52 weeks **4.** 14

Friday Favorite Breakfast Food chart Tally column: Cereal 𝍸; Eggs 𝍸 𝍸; Pancakes 𝍸 I; Granola 𝍸 I

1. eggs **2.** 11 **3.** pancakes and granola **4.** cereal

Brain Stretch **1.** 94 **2.** 86 **3.** 63 **4.** 42

Week 15, pages 43–45

Monday **1.** ≠ **2.** 8 **3.** 5 **4.** 7 + 12 = 19; 12

Tuesday **1.** □□□□ IIIIIII ••••••••• **2. A.** 70 **B.** 300 **3. A.** even **B.** odd **4.** 40¢

Wednesday **1.** pyramid **2.** 8 **3.** 5 **4.** ⬠ **5.** B

Thursday **1.** B **2.** 51; 51 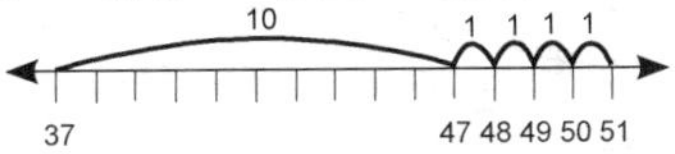

3. 365 days **4.** Lines should measure 5 cm.

Friday **1.** juice **2.** water **3.** 36 **4.** lemonade and milk

Brain Stretch **1.** 82 **2.** 91 **3.** 15 **4.** 2

Week 16, pages 46–48

Monday **1.** ≠ **2.** 7 **3.** 4 **4.** 3 + 3 + 3 + 3 + 3 = 15

Tuesday **1.** 5 hundreds, 4 tens, 2 ones, number 542 **2.** 73 > 21 > 17 **3.** 3 + 5 = 8, 8 – 3 = 5 or 8 – 5 = 3 **4.** One side of the shape should be colored.

Wednesday **1.** The 3 triangles should be colored. **2.** ⬡ **3.** A

Thursday **1.** 6:15, quarter past 6 **2.** A **3.** A **4.** 9

Friday **1.** Those on the left side: Ben, Alma, Tiffany, Victoria **2.** Those on the right side: Sylvia, Paula, Arthur, Juan, Cory, Maria **3.** Those in the middle: Katie and Mike

Brain Stretch **1.** 84 **2.** 90 **3.** 26 **4.** 16

Week 17, pages 49–51

Monday **1.** = **2.** 7 **3.** 2 **4.** 620

Tuesday **1.** □□ ||||||||| ••••• **2.** 5 + 7 = 12, 12 – 7 = 5 or 12 – 5 = 7 **3. A.** 812 **B.** 541 **4.** One section of the shape should be colored.

Wednesday **1.** The first, third, and fifth pentagons should be colored. **2.** ⬡ **3.** B

Thursday **1.** 3:45, quarter to 4 **2.** B **3.** A **4.** 13

Friday **1.** Those on the far left: Jessie, Avita, Carlos, Jeffery, Alice, Sandra **2.** Those in the right circle: Linda, Leah, Amanda, Dana, James, John **3.** Those on the far right: Leah, Amanda, Dana, James, John

Brain Stretch **1.** 95 **2.** 75 **3.** 28 **4.** 29

Week 18, pages 52–54

Monday **1.** = **2.** repeating **3.** Accept any number pattern with matching rule. **4.** 450

Tuesday **1. A.** 70 **B.** 300 **2. A.** 711 **B.** 344 **3.** 6 tens, 3 ones **4.** One hippopotamus should be circled.

Wednesday **1.** Colors, in order: red, yellow, blue, green **2.** A **3.** A

Thursday **1.** The short hand should point to 11 and the long hand to 9. **2.** 4; 4 inches **3.** B **4.** 10

Friday **1.** Those on the far left: Jacob, Isabella, Noah **2.** Those on the far right: William, Jimmy, Grace **3.** 3

Brain Stretch **1.** 905 **2.** 878 **3.** 237 **4.** 366

Week 19, pages 55–57

Monday **1.** 20 **2.** 25 **3.** growing **4.** 32

Tuesday **1. A.** 8 **B.** 90 **2.** 9 + 7 = 16 or 7 + 9 = 16, 16 – 9 = 7 **3. A.** 400 + 30 + 8 **B.** 200 + 90 + 5 **4.** One side of the shape should be colored.

Wednesday **1.** Colors, in order: orange, blue, red, and green **2.** A **3.** Sample answer: ⧄

Thursday **1.** 9:15, quarter past 9 **2.** B **3.** 37 – 17 = 19; 19

Friday **1.** Those on the far left: Sophie, Chris, Gina **2.** 6 **3.** Those in the middle: David, Demetra, Andrew

Brain Stretch **1.** 959 **2.** 996 **3.** 172 **4.** 349

Week 20, pages 58–60

Monday **1.** 10 **2.** 30 **3.** shrinking **4.** 626

Tuesday **1.** C **2.** 8 + 5 = 13 or 5 + 8 = 13, 13 – 8 = 5 **3. A.** odd **B.** odd **4.** One section of the shape should be colored.

Wednesday **1.** Colors, in order: green, blue, red, orange **2.** B **3.** △

Thursday **1.** 5:15, quarter past 5 **2.** 2 cups **3.** 44 –18 = 26; 26 inches **4.** A

Friday **1.** 31 **2.** Saturday **3.** Monday **4.** Thursday

Brain Stretch 27¢

Week 21, pages 61–63

Monday **1.** 11 **2.** 9 + 11 = 20; 20 **3.** Accept any pattern with a matching rule.

Tuesday **1.** B **2.** B **3.** 419 + 135 = 554 (number line: 419 → +100 → 519 → +10 → 529 → +10 → 539 → +10 → 549 → +5 → 554) **4.** One penguin should be circled.

Wednesday **1.** The right 3 shapes should be colored: hexagon, square, pentagon **2.** B **3.** Sample answer:

Thursday **1.** 10:45, quarter to 11 **2.** 3 hours **3.** A **4.** 16 ounces

Friday **1.** 30 **2.** Friday **3.** Tuesday **4.** Saturday

Brain Stretch 14 pets

Week 22, pages 64–66

Monday **1.** 3 **2.** 15 **3.** Accept any alternating pattern.

Tuesday **1.** A **2.** 881 **3.** 673 **4.** One turtle should be circled.

Wednesday **1.** The hexagon and octagon should be colored. **2.** A **3.** Accept any line that creates two equal halves.

Thursday **1.** 11:15, quarter past 11 **2.** 100 cm **3.** 6 squares **4.** 8 squares

Friday **1.** favorite meals **2.** dinner **3.** lunch **4.** dinner

Brain Stretch 18 frogs

Week 23, pages 67–69

Monday **1.** ≠ **2.** 11 **3.** Accept any pattern with a matching rule. **4. A.** 338 **B.** 318

Tuesday **1.** B **2.** Yes. 40 + 20 = 60; 5 + 5 = 10; 60 + 10 = 70; So, 45 + 25 = 70. **3. A.** 500 **B.** 700 **4.** 5 + 8 = 13, 13 – 5 = 8 or 13 – 8 =5

Wednesday **1.** All shapes but the triangle should be colored. **2.** B **3.**

Thursday **1.** 8:45, quarter to 9 **2.** B **3.** 6 squares **4.** 9 squares

Friday **1.** apples **2.** grapes **3.** grapes < bananas < apples **4.** 11 **5.** 6

Brain Stretch 92 jelly beans

Week 24, pages 70–72

Monday **1.** 31 **2.** 3 + 12 + 4 = 19; 19 **3.** 59 **4.** Accept any pattern with a matching rule.

Tuesday **1.** B **2.** 41; 25 + 16 = 41 **3. A.** 730 **B.** 204 **C.** 511 **4.** 3 robots should be circled.

Wednesday **1.** All but the octagon and the hexagon should be colored. **2.** A **3.** B

Thursday **1.** 10:00 **2.** A **3.** 4; 4; fourth **4.** 12 squares

Friday **1.** pop **2.** country **3.** pop > rock > country **4.** 22 **5.** 12

Brain Stretch 48 hockey cards

Week 25, pages 73–75

Monday **1.** 42 **2.** 76 **3.** 560, 570, 580, 590, 600 **4.** 124

Tuesday **1.** 234 **2.** 390, 285, 176 **3. A.** 500 **B.** 90 **4.**

Wednesday **1.** In order, the objects should be colored: blue, yellow, green, red, yellow, blue, red **2.** A **3.**

Thursday **1.** The short hand should point between 10 and 11, and the long hand should point to the 5. **2.** B **3.** 15 squares **4.** 2; half

Friday The bar graph should be shaded to 6 for Art, 9 for Language, 8 for Math, 8 for Science, and 4 for Music. **1.** language **2.** 14 **3.** Math and science.

Brain Stretch 60 cookies

Week 26, pages 76–78

Monday 1. 322, 422, 522, 622 2. 26 3. 17 – 9 = 8; 8 4. 537

Tuesday 1. 149 2. 2 + 2 + 2 + 2 + 2 + 2 = 12 3. A. 100 B. 200 4.

Wednesday 1. C 2. A 3. B 4. Sample answer:

Thursday 1. 2:45, quarter to 3 2. B 3. B 4. 8

Friday 1. Ferris wheel 2. The roller coaster and the merry-go-round. 3. 4 4. 8

Brain Stretch 24 ants

Week 27, pages 79–81

Monday 1. 890, 895, 900, 905 2. 52 3. 575, 580, 585, 590, 595 4. ≠

Tuesday 1. 5 + 5 = 10 or 2 + 2 + 2 + 2 + 2 = 10 2. B 3. A. 400 B. 200 4. 9 + 5 = 14 or 5 + 9 = 14, 14 – 9 = 5

Wednesday 1. Color of shapes, in order: blue, red, blue, yellow, green, red, yellow 2. B 3. B

Thursday 1. The short hand should point between 2 and 3, and the long hand should point to 8.
2. 3; third 3. Lines should be 5 inches long. 4. 2; half

Friday The Favorite Color numbers should be: 9 Red, 13 Blue, 8 Green, and 2 Yellow
1. blue 2. yellow 3. 10 4. blue > red > green > yellow

Brain Stretch 10 nickels

Week 28, pages 82–84

Monday 1. 51 2. 68 3. A. 145 B. 358 C. 551 4. 14

Tuesday 1. 5 + 5 + 5 + 5 = 20 *or* 4 + 4 + 4 + 4 + 4 = 20 2. A 3. 9 + 3 = 12, 12 – 9 = 3 or 12 – 3 = 9 4. Three sections of the shape should be colored.

Wednesday 1. Colors, in order: red, blue, green, orange, green, red, orange 2. A or B 3.

Thursday 1. 9:00 2. Accept any answer 70°F or higher. 3. 95 cents 4. A

Friday 1. Those on the far right: Stephen, Monique, Chloe 2. Those in the left circle: Chris, Bess, Ethan
3. Those in the middle: Bess, Ethan

Brain Stretch 12 cupcakes

Week 29, pages 85–87

Monday 1. 30 2. 33 3. Accept any pattern with a matching rule. 4. 23 + 14 = 37 *or* 37 – 14 = 23; 23

Tuesday 1. 3 + 3 + 3 + 3 + 3 = 15 or 5 + 5 + 5 = 15 2. A. 486 B. 912 C. 102 3. A. 600 B. 700 4. 150

Wednesday 1. The 3 triangles should be colored. 2. C 3. 4

Thursday 1. 1:30 2. 2 quarts 3. 55 cents 4. B

Friday Favorite Transportation chart Tally column: Skateboard 卌 ||||; Bike 卌 卌 卌 ||; Scooter 卌 卌 |||
1. bike 2. skateboard 3. 8

Brain Stretch 8¢

Week 30, pages 88–90

Monday 1. 15 2. 20 3. 23; 23

Tuesday 1. 3 + 3 = 6 *or* 2 + 2 + 2 = 6 2. A. 97 B. 560 C. 711 3. A. 246 B. 12 C. 66 4. 153

Wednesday 1. The 3 circles should be colored. 2. A or C 3. 5

Thursday 1. 11:45, quarter to 12 2. Lines should be 3 inches long. 3. 45 cents 4. 7 3; one-third

Friday 1. Friday 2. Monday 3. 8

Brain Stretch Sample answer: 4 quarters, 10 dimes, 100 pennies

Common Core State Standards for Mathematics Grade 2

Student	2.OA.1	2.OA.2	2.OA.3	2.OA.4	2.NBT.1	2.NBT.2	2.NBT.3	2.NBT.4	2.NBT.5	2.NBT.6	2.NBT.7	2.NBT.8	2.NBT.9	2.MD.1	2.MD.2	2.MD.3	2.MD.4	2.MD.5	2.MD.6	2.MD.7	2.MD.8	2.MD.9	2.MD.10	2.G.1	2.G.2	2.G.3

Level 1 Student demonstrates limited comprehension of the math concept when applying math skills.

Level 2 Student demonstrates adequate comprehension of the math concept when applying math skills.

Level 3 Student demonstrates proficient comprehension of the math concept when applying math skills.

Level 4 Student demonstrates thorough comprehension of the math concept when applying math skills.

Math — Show What You Know!

☐ I read the question and I know what I need to find.

☐ I drew a picture or a diagram to help solve the question.

☐ I showed all the steps in solving the question.

☐ I used math language to explain my thinking.